AF874999

Understanding Complex Systems

Founding Editor

Prof. Dr. J.A. Scott Kelso
Center for Complex Systems & Brain Sciences
Florida Atlantic University
Boca Raton FL, USA
E-mail: kelso@walt.ccs.fau.edu

Editorial and Programme Advisory Board

Dan Braha
New England Complex Systems, Institute and University of Massachusetts, Dartmouth

Péter Érdi
Center for Complex Systems Studies, Kalamazoo College, USA and Hungarian Academy of Sciences, Budapest, Hungary

Karl Friston
Institute of Cognitive Neuroscience, University College London, London, UK

Hermann Haken
Center of Synergetics, University of Stuttgart, Stuttgart, Germany

Viktor Jirsa
Centre National de la Recherche Scientifique (CNRS), Université de la Méditerranée, Marseille, France

Janusz Kacprzyk
System Research, Polish Academy of Sciences, Warsaw, Poland

Kunihiko Kaneko
Research Center for Complex Systems Biology, The University of Tokyo, Tokyo, Japan

Scott Kelso
Center for Complex Systems and Brain Sciences, Florida Atlantic University, Boca Raton, USA

Markus Kirkilionis
Mathematics Institute and Centre for Complex Systems, University of Warwick, Coventry, UK

Jürgen Kurths
Potsdam Institute for Climate Impact Research (PIK), Potsdam, Germany

Linda Reichl
Center for Complex Quantum Systems, University of Texas, Austin, USA

Peter Schuster
Theoretical Chemistry and Structural Biology, University of Vienna, Vienna, Austria

Frank Schweitzer
System Design, ETH Zürich, Zürich, Switzerland

Didier Sornette
Entrepreneurial Risk, ETH Zürich, Zürich, Switzerland

For further volumes:
http://www.springer.com/series/5394

Understanding Complex Systems

Future scientific and technological developments in many fields will necessarily depend upon coming to grips with complex systems. Such systems are complex in both their composition - typically many different kinds of components interacting simultaneously and nonlinearly with each other and their environments on multiple levels - and in the rich diversity of behavior of which they are capable.

The Springer Series in Understanding Complex Systems series (UCS) promotes new strategies and paradigms for understanding and realizing applications of complex systems research in a wide variety of fields and endeavors. UCS is explicitly transdisciplinary. It has three main goals: First, to elaborate the concepts, methods and tools of complex systems at all levels of description and in all scientific fields, especially newly emerging areas within the life, social, behavioral, economic, neuroand cognitive sciences (and derivatives thereof); second, to encourage novel applications of these ideas in various fields of engineering and computation such as robotics, nano-technology and informatics; third, to provide a single forum within which commonalities and differences in the workings of complex systems may be discerned, hence leading to deeper insight and understanding.

UCS will publish monographs, lecture notes and selected edited contributions aimed at communicating new findings to a large multidisciplinary audience.

Springer Complexity

Springer Complexity is an interdisciplinary program publishing the best research and academic-level teaching on both fundamental and applied aspects of complex systems - cutting across all traditional disciplines of the natural and life sciences, engineering, economics, medicine, neuroscience, social and computer science.

Complex Systems are systems that comprise many interacting parts with the ability to generate a new quality of macroscopic collective behavior the manifestations of which are the spontaneous formation of distinctive temporal, spatial or functional structures. Models of such systems can be successfully mapped onto quite diverse "real-life" situations like the climate, the coherent emission of light from lasers, chemical reaction-diffusion systems, biological cellular networks, the dynamics of stock markets and of the internet, earthquake statistics and prediction, freeway traffic, the human brain, or the formation of opinions in social systems, to name just some of the popular applications.

Although their scope and methodologies overlap somewhat, one can distinguish the following main concepts and tools: self-organization, nonlinear dynamics, synergetics, turbulence, dynamical systems, catastrophes, instabilities, stochastic processes, chaos, graphs and networks, cellular automata, adaptive systems, genetic algorithms and computational intelligence.

The two major book publication platforms of the Springer Complexity program are the monograph series "Understanding Complex Systems" focusing on the various applications of complexity, and the "Springer Series in Synergetics", which is devoted to the quantitative theoretical and methodological foundations. In addition to the books in these two core series, the program also incorporates individual titles ranging from textbooks to major reference works.

Octavian Iordache

Self-Evolvable Systems

Machine Learning in Social Media

Springer

Author
Octavian Iordache
Polystochastic
Montreal
Quebec
Canada

ISSN 1860-0832 e-ISSN 1860-0840
ISBN 978-3-642-28881-4 e-ISBN 978-3-642-28882-1
DOI 10.1007/978-3-642-28882-1
Springer Heidelberg New York Dordrecht London

Library of Congress Control Number: 2012935114

© Springer-Verlag Berlin Heidelberg 2012
This work is subject to copyright. All rights are reserved by the Publisher, whether the whole or part of the material is concerned, specifically the rights of translation, reprinting, reuse of illustrations, recitation, broadcasting, reproduction on microfilms or in any other physical way, and transmission or information storage and retrieval, electronic adaptation, computer software, or by similar or dissimilar methodology now known or hereafter developed. Exempted from this legal reservation are brief excerpts in connection with reviews or scholarly analysis or material supplied specifically for the purpose of being entered and executed on a computer system, for exclusive use by the purchaser of the work. Duplication of this publication or parts thereof is permitted only under the provisions of the Copyright Law of the Publisher's location, in its current version, and permission for use must always be obtained from Springer. Permissions for use may be obtained through RightsLink at the Copyright Clearance Center. Violations are liable to prosecution under the respective Copyright Law.
The use of general descriptive names, registered names, trademarks, service marks, etc. in this publication does not imply, even in the absence of a specific statement, that such names are exempt from the relevant protective laws and regulations and therefore free for general use.
While the advice and information in this book are believed to be true and accurate at the date of publication, neither the authors nor the editors nor the publisher can accept any legal responsibility for any errors or omissions that may be made. The publisher makes no warranty, express or implied, with respect to the material contained herein.

Printed on acid-free paper

Springer is part of Springer Science+Business Media (www.springer.com)

Gignit autem artificiosam lusorum gentem Cella Silvestris

But Waldzell breeds the skillful Glass Bead Game players

Hermann Hesse

The Glass Bead Game, Zürich, 1943

Preface

Chemical engineering schemas, biochemical reactions networks, large scale software and hardware systems, organisms, companies, supply networks, markets, ecosystems, all are examples of complex systems. Complexity research tries to identify general principles of emerging organizations common to such systems across diverse areas, to understand their organizational structure in a coherent and rigorous way.

Significant contemporary concerns as, accessibility to energy, food, drugs and health care, environmental degradation and market uncertainty pertain without doubt to evergrowing complexity sphere. We lack the scientific tools to think clearly about these problems and to confront emergent phenomena.

A challenging research direction in high complexity domains is the pursuit of autonomy for systems. However, it was observed that existing automatic systems as bio-chemical or nuclear reactors, personalized medical devices, environment, market or security controllers, will inevitably fail to meet some critical tasks or obligations if they are not self-evolvable. The self-evolvability refers to systems capable of self-building, self-organizing, self-managing, self-repairing and encountering several other aspects of self-properties at the level of taking care of themselves.

Self-evolvability is advanced here as a key strategy to confront the evergrowing complexity of systems.

The self-evolvability is the response to the shift from needing faster and stronger hardware integrated computer systems, to the need for less external-intensive management of those systems. Certainly, the systems complexity will reach the point where external administration and control is impossible or will cost much more than the integrated hardware and software infrastructure. The fact that external administrators get more and more overstrained by evolution tasks have lead to the idea of self-evolving systems.

Self-evolvability incorporates specific theoretical concepts and methods as the categorification, decategorification and self-integrative closure since it needs modeling at different abstraction levels and this is deep-rooted in the n-categories study. It is the evergrowing complexity that imposes to ascend and descend the steps of the higher dimensional categories in modeling.

Self-evolvability includes technological aspects. This means, for the industrial systems case study, that the production should be flexible and the new modules should be flexibly integrated with existing materials, equipments or programs that are still reusable, reconfigurable and evolvable.

Self-evolvability encompasses financial aspects as making balanced and durable investments in a way that will not be wasted when the market or context evolves, and socio-economic features as the strategies to avoid complete off-shoring of tasks or entire industries.

This book envisages describing systems that will be able to combine technological, scientific and engineering, aspects of complexity into quintessential expressions as the paradigm of self-evolvability.

Emphasizing the multiple features involved by self-evolvability, the attention here is focused on innovative concepts and models, and also on relevant applicative domains in which they can be efficiently used.

The book is divided in 12 chapters.

The first chapter introduces the self-evolvability as the general strategy to face and succeed the evergrowing complexity.

Levels of reality and categories, modeling architectures, polystochastic models, and self-integrative closure are presented in chapter 2.

Chapter 3 introduces differential models and appropriate dynamical concepts for multiple conditioning levels. New types of models, formally similar to the classical ones that will allow now to design, schemas, experiment designs and architectures are introduced.

New informational entropy criteria are presented in the fourth chapter.

Chapters 5 to 11 provide case studies. Chapter 5 outlines self-evolvability for physical systems as chemical engineering separation schemas, dendritic growth and electrical circuits. Chapter 6 analyzes biochemical models based on polytopic architectures and correlates these with biological and bio-inspired computing models.

The closure and post-formal aspects of cognitive developmental stages, logical hypercube, and relational complexity are presented in chapter 7. Based on these, conceptual and computational frameworks for intelligent and autonomous cognitive architectures are proposed.

Chapter 8 focuses on control systems. The emergence of self-control capability for high dimensional automata, self-control architectures and different types of interconnections are discussed.

Chapter 9 is dedicated to viable enterprises and self-evolvable manufacturing systems.

Chapter 10 studies self-evolvability aspects for concept lattices as encountered in formal concept analysis, and applied for biochemical computations and hierarchical classes analysis.

Chapter 11 evaluates the model generated designs of experiments and the self-evolvability potential of designs with applications to pharmaceutical systems and drugs design, to quality evaluations for circuits.

Chapter 12 discusses future researches including the selfdisciplinarity and the polytope project. This project refers to a biologically inspired overarching framework shared by the operational structure of self-evolvable devices, the functional organization of organisms as informational and cognitive systems, and the scientific and engineering methods.

Appendices introduce elements of n-category theory emphasizing on operads, rewriting and polycategories concepts.

Self-evolvability solutions to high complexity should be envisaged giving that classical engineering devices, tools, methodologies or organizations reached or will reach their limits and alternatives ways of solving problems are now required.

Numerous self-evolvability studies are now in progress in different domains of activity but general frameworks for self-evolvability systems are still missing and should be created. The case studies sketched in the book may represent a source of inspiration for emerging methods, technologies and systems in their mandatory transition from adaptable and evolvable towards self-evolvable. The presented case studies show what we need to understand to build high complexity systems of their own and better understand and manage those existing in the world.

Introducing a new field of major practical and theoretical interest and a key area for future investigations of evergrowing complexity, the book will be useful to engineers, scientists, entrepreneurs and students in different domains of engineering, systems science, and applied mathematics.

Contents

List of Figures

List of Tables

Abbreviations

ANOVA	Analysis of Variance
CT	Category Theory
DNA	Deoxyribonucleic Acid
DOE	Design of Experiment
EDOE	Evolvable Design of Experiment
FCA	Formal Concept Analysis
GL	Galois Lattice
HCA	Hierarchical Classes Analysis
IT	Information Technology
LIDA	Learning Intelligent Distribution Agent
LISA	Learning and Inference with Schemas and Analogies
OLAP	On Line Analytical Processing
PSM	Polystochastic Model
RNA	Ribonucleic Acid
SASO	Self-Adaptive and Self-Organizing
SKUP	States, Conditions, Operators, Possibilities
STAR	Structured Tensor Analogical Reasoning
WE	Wave Equation

Chapter 1
Introduction

Abstract. Self-evolvability is advanced here as the key method to successfully manage the evergrowing complexity of systems.

The necessary transition from adaptable, to evolvable and finally to self-evolvable systems is highlighted.

Self-properties as self-organization, self-configuration, self-repairing and so on have been briefly introduced.

Challenges and limitations of the self-evolvable engineering systems are evaluated.

1.1 Self-Evolvable Systems to Manage Complexity

Technical systems as industrial equipments, telecommunication networks, and socio-technical systems as manufactures, companies and societies became more and more complex. This is the result of natural increasing complexity for products and markets, of the embedding of hardware, software, logical tools and necessary knowledge into such systems. The evergrowing complexity of modern industrial systems will lead to unsustainable increases in their management and operation costs.

Problems of design and organization become pressing as products increase in complexity regarding both hardware and software. It is becoming inevitable to shift much of the problem of organization into the machines themselves. This brings up the problem of keeping the self-organization under control. Product, installations, hardware and software engineers aim to relieve growing complexity problems by conceiving adaptable, evolvable and finally self-evolvable systems.

A self-evolvable system should be capable to adapt dynamically to the current conditions and future requests of its environment.

Conventional engineering or scientific methods have been conceived to deal with systems reducible to simpler parts which exhibit controlled behavior.

A significant step in high complexity management was and continues to be the pursuit of autonomy for systems. However, the autonomous robotic systems will inevitably fail to meet some tasks or obligations when they are operated in different field conditions, if they are not self-evolvable.

Self-evolvability is advanced here as the key method to confront evergrowing complexity and to successfully run in the higher complexity domains.

The self-evolvability paradigm is the proposed response to the shift to faster and stronger hardware integrated computer systems, to the need for less external management of high complexity systems.

To explain the self-evolvability need, we start from the observation that there are significant differences between adaptable, evolvable and self-evolvable systems.

Fig. 1.1 illustrates the trends and steps in understanding and system modeling as complexity grows. Adaptability, based on learning, implies optimization or adjustment on the time scale of the existence of a system as for instance an industrial product or organization. Adaptability may refer to animate and its inanimate environment. It offers a preliminary and low dimensional perspective for complexity running.

In a simplified form, the learning models describing adaptability considered two spaces, the space of states, S and the space of conditions, K and their interaction (Fig. 1.1a). The learning system is presented with a series of conditions $k \in K$, on each of which it changes states $s \in S$, allowing sequential adaptability (Bush and Mosteller 1955, Iosifescu and Grigorescu 1990).

The evolvability requires more than learning, specifically the capacity for change to autonomous march into new life cycles, for instance new type of products, new market niches, new organizations and new levels. This entails a higher dimensional perspective and also the systems closure.

The evolvability refers to several levels, for instance, to a transformation from physical states S, toward biological-like, K1, cognitive-like, K2, intelligent-like, K3 and finally closed evolvable systems. Four levels systems in which the states S are related to a hierarchy of conditions K1, K2, K3 have been studied in different engineering domains where evolvability was implemented (Iordache 2010).

Fig 1.1b shows the basic cognitive frame for evolvability in which the states S interact with the conditions K1, meta-conditions K2, and meta-meta conditions K3.

Critical to achieve evolvability is the embodiment that is the connection between the top level of conditions K3 and the states S. Artificially or naturally evolvable systems are supposed to cross this critical gap between these two extreme levels.

The continuous growth of complexity imposes a new transition from evolvability to self-evolvability, that is to systems that self-configure, self-optimize, self-control, self-heal and so on, based on a set of higher-level intrinsic capabilities and meeting of the user-specified variable objectives.

The polytope shown in Fig. 1.1c illustrates the architecture of such self-evolvable systems. It is a representation known as hypercube or 4-cube and consists of a cube inside another cube. The 4-cube is obtained by joining all corners of the inner cube with the corresponding corners of the outer cube (Ziegler 1995).

As shown in Fig. 1.1c, complementing the direct way of integration and convergence $S \rightarrow K1 \rightarrow K2 \rightarrow K3$ we need to look at the reverse way of differentiation and divergence $K3' \rightarrow K2' \rightarrow K1' \rightarrow S'$.

The convergence way does not quite grasp the essence of creativity required by self-evolvability. That is because the boundaries where creativity flourishes and new information and new solutions are created consist of synchronized tendencies. Tendencies to converge should coexist with tendencies to diverge and it is the metastable blend of both that matters (Kelso 2002).

The Self-cube centers and correlates the four-level evolvable frames shown on different faces of the outer cube (Fig. 1.1c). Swinging, mediated by the Self between the two complementary way's is crucial for self-evolvability.

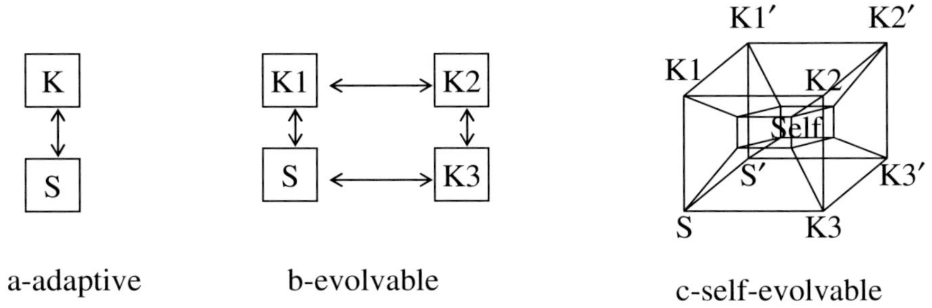

Fig. 1.1 Modeling architectures for growing complexity

Evergrowing complexity management needs conveying, as shown in Fig. 1.1, from adaptive, to evolvable and lastly to self-evolvable systems. The self-evolvability subsumes and challenges the adaptability and the evolvability stages.

For the design and control of forthcoming self-evolvable systems a natural query is: why not use traditional methods, based on modeling and extrinsic implementation of the models in the usual computer based adaptive design and control?

A two-level relation between states S and conditions K, as shown in Fig. 1.1a, may accomplish some adaptive control tasks. But the imagined strategy is not to reduce several interconnection steps to the two levels architecture, as an attempt to achieve a degree of control in the speediest manner. Contrary to this kind of reductionism, the evolvability schemas shown in Fig. 1.1b look to the basic four interactions steps to allow the study and to take into account the necessary ingredients, as shown by the evolvable systems and organisms existing in nature.

From the computer science and engineering points of view, the apparent ease with which living systems solve computationally difficult organizational problems makes it inevitable to adopt the strategies observed in nature for creating information processing architectures.

For systems of higher complexity, the envisaged computing and control tasks are impossible to be extrinsically operated. For conventional design and control the majority or non-linear interactions that could contribute to the problem are inherently excluded. The properties characterizing high complexity systems should be the consequences of their own dynamic of the computational environment, not of the decision of the external designer or program that is anyway unable to predict the evolution of its construction.

Self-evolvable systems are intended to work for their building and evolution more efficiently than any external computer-aided operator can do. Only self-evolvable systems will have the potential to survive in unforeseen situations.

Additionally, it was observed that the more an adaptive system is optimized to perform a task the less capability it has to deal with unexpected or unprogrammed major changes in an uncertain environment. This implies that for highly complex environments, self-evolvability rather than optimization or adaptability may be the appropriate measure of the system potential to carry out high complexity tasks and to survive.

Building on the adaptable and evolvable systems, we see the potential for the development of sophisticated methods, devices and controllers for use on autonomous systems. But it is impossible for engineers to anticipate all possible events and non-linear combinations of events that may be faced by systems in hazardous, hostile, or increasingly complex environments. Providing the industrial system with a representation of human-like cognition and a knowledge base from which decisions may be developed will greatly expand the applicability of autonomous systems as solutions to high complexity problems (Strunk and Ganger 2003).

It is acknowledged that as systems continue to grow in complexity, they will reach a point where administrators will not only be unable to understand the behavior but will lack the ability to control that system by conventional methodologies. A breakthrough is necessary since research is at the crossroads and the conventional methods will no longer work (Ritchey 2011).

An embedded human-like cognitive control capability may provide the means for establishing reliable control of these systems. This is the perspective visualized by the self-evolvable architectures from Fig. 1.1c. This outlines the Self, correlating the two ways of integration and differentiation.

Since there is no fixed limit for growing complexity, higher dimensional polytopes as 5-cubes and other polytopes will be considered as cognitive architecture instead of the 4-cube shown in Fig. 1.1c. Self-evolvability is the growth toward greater understanding and control not its attainment.

At the today technological level, a project grouping in a system the major faculties of self-evolvability becomes realistic.

Undoubtedly the agenda for development of self-evolvable systems requires a transdisciplinary and selfdisciplinary effort. It is expected to implement self-evolvability as an increasing capability in steps. Gradually it will be possible to make a system more close to self-evolvable ones as the complexity understanding and resources to prevail over high complexity will grow.

Self-evolvability will emerge as a natural consequence of step-by-step implementations of the self-properties.

1.2 Self-Properties

The term self-property generally refers to the acquirement of the indicated property in the absence of external intervention or control.

The self-evolvable systems are based on integrated hardware, software and selfware infrastructure. Selfware concept defines the growing set of self-properties that are emerging in self-managing computing systems major initiatives as for instance autonomic computing or organic computing (Sterritt and Hintchey 2005).

The initial set of self-properties, namely self-configuration, self-healing, self-optimization and self-protection as objectives to be attained through self-awareness, self-monitoring and self-adjusting attributes, has been expanded, and further properties are expected to be added to this still growing list.

Diverse interpretations of concepts as self-properties and emergence can be found in the literature (Banzhaf 2002a, 2002b, Frei and Di Marzo Serugendo 2011a, 2011b).

A taxonomy of self-properties which focuses on decentralized autonomic computing and discusses characteristics of self-properties and implications for their engineering was proposed (De Wolf and Holvoet 2007). This taxonomy takes into account if a self-property is achieved on macroscopic or microscopic level, if it is on-going or one-shot, if it is time dependent or independent, if it evolves in a continuous or abrupt way, and it is adaptation-related or not. The taxonomy gives examples of mechanisms leading to self-properties and classifies application examples according to the considered characteristics.

The self-properties of interest, mainly for industrial engineering systems, will be briefly characterized in the following.

• Self-organization

Self-organization is defined as a process where systems acquire and maintain structures themselves, without external control. Self-organization can be broadly understood as the ability of a system to change its internal structure and its function in response to external circumstances.

A system creates or adapts its own structure to reach a goal. Components may form and break coalitions to provide the requested capabilities. A self-organizing system can assemble, construct and stabilize itself, with the help of outside matter, energy or information.

Self-organization would not be possible without nonlinear interactions between components of the system. The self-organizing systems are open-ended and not in static equilibrium state with their environment. Self-organization includes the increase in structure, autonomy and robustness with reference to changes and far-from-equilibrium conditions.

• Self-adaptation

A self-adaptive system adjusts itself to changing conditions without major physical modifications. For instance, in the case of an industrial production system when more urgent requests arrive, an automaton can modify its working parameters (Frei 2010).

Self-adaptation is one of the processes by which a system can self-organize. Whether it is a system of brain cells that adapt to firing patterns, or a system of street signals that adapt to traffic, or a system of product manufacturing cells that adapt to new product requests, similar formal models might be able to describe those systems.

- Self-configuration and Self-reconfiguration

These refer to systems that automatically configure and reconfigure components to adapt them to different environments. Self-configuring is based on feedback from environment.

A self-configuring system prepares itself for functioning, including the adjustment of parameters and calibration. Automata adjust their geometry and movement accuracy to the desired range of requests (Frei and Di Marzo Serugendo 2011a, 2011b).

Self-reconfiguration encompasses self-adaptation, but also basic change, including the software and hardware. For instance, when a component or a cell fails, and there is an alternative component or cell path to reach the affected destination, the parts adapt their behavior and use the alternative path until the part has recovered from the failure. A new device or cell is required and it should be integrated into the existing system.

A self-reconfigurable modular approach was imagined to build responsive industrial systems to satisfy customers, various demands (Mun et al. 2004, Hu and Efstathiou 2007).

- Self-assembly and Self-disassembly

The self-assembly refers to sub-systems or parts that connect with each other, to form the totality.

By disassembly, system decomposes itself into subsystems or parts. An association which is not necessary any more may disassembles or may be disconnected.

The disassembly process allowing the formulation of new systems is as important as the self-assembly itself.

There are some differences between self-assembly, self-formation, and self-organization concepts (Banzhaf 2002a).

Self-assembly comprises the assembly of parts into a whole, directed by the assembling parts and their interactions. A self-assembling process is usually not recursive, that is, it cannot move through successive stages of first assembling some elementary parts into more complex parts, which in turn self-assemble into the whole.

Self-formation has a clear feature of sequence. Processes of self-formation can be used to generate more complicated entirety. This requires the developing system to change state repeatedly, with each state determining subsequent states and the sequence of events to follow. More complex wholes can be constructed by such a mechanism, and it is possible to generate patterns of higher complexity. Self-formation finds its limits in the problem of repairing and maintenance of the structures formed. Because of the very specific sequence of events that lead to the original result, these needs are difficult to achieve with self-formation.

We may consider self-organization to be the most general term in this order, including both self-assembly and self-formation, but also self-maintenance and self-development.

Self-organization does not require a specific sequence to arrive at the end result since it has a multitude of paths toward the desired goal. Self-organization allows phenomena on different time scales, and a hierarchy of levels, which in turn allows a recursive consideration of its mechanisms.

• Self-diagnosis, Self-repairing and Self-healing

Components can find out and state what is not right with their functioning. A device that cannot provide products may check if there are no ready products inside, or if there is a blockage or any other problem preventing normal performance (Barata et al. 2007).

Self-repairing refers to systems that automatically discover, diagnose and correct faults.

A system can treat its problems and maintain or re-establish functionality. A blocked path or device will restart its software, execute calibration movements, check and if still blocked, may ask the user for debugging. Self-healing implies a level of tolerance to errors.

• Self-reproduction, Self-replication and Self-fabrication

These properties refer to systems that can create copies. A modular assemblage supports suitable modules to form the same type of assemblage. Self-similarity may be a feature of the resulting system.

The self-fabrication refers to fabrication process that happens spontaneously without assistance from a fabricator. Hofmeyr identifies unassisted self-assembly as the process that ultimately makes the system self-fabricating (Hofmeyr 2007).

Hofmeyr characterizes living systems as featuring persistence despite a continual decay of their molecular components and machinery and despite changes in context. This led to present a living system as a chemical factory of which the output is the factory itself.

The concept of autonomous self-fabrication was developed in the theory of autopoietic systems (Maturana and Varela 1980) and of self-reproducing automata based on universal constructors (Von Neumann 1966).

• Self-protection and Self-security

These refer to systems that anticipate, identify and protect against arbitrary attacks.

A self-evolvable system should protect itself from interferences or attacks. In case a system was open enough for external agents to gain access to it, it would need to protect itself from possible damages. A self-protecting system observes, constructs, knocks down and modifies its boundaries.

• Self-control, Self-determination and Self-management

These refer to a system that guides itself. The modules control their own behavior, for instance that directed by rules and policies. Self-management should characterize a system that can take care of itself. This may include self-protection, self-healing, self-configuration, self-optimization, and self-adaptation. Self-optimization, for instance, means that the system automatically monitors and adapts resources to ensure optimal functioning regarding the defined requirements.

For example, at production time, the processing components should maintain themselves as well as their neighbors in safe conditions. They should manage their multi-lateral interactions providing the requested services and schedules maintenance.

• Self-awareness

This refers to the capability of the system or its individual components to identify by themselves, internally, any new condition, failure, or problem, without specifically being instructed, from outside, by any administrator. A system is self-aware in that it can observe itself and improve its behavior to meets its goals.

Self-awareness requires sensing capabilities and triggers reasoning and acting.

Self-aware systems are currently thought of as equipped with monitoring, planning and plan execution capabilities at the level of autonomic managing. Self-aware systems sense their environment in different ways as configurations or neighbors, and take decisions accordingly, changing directions, roles, goals or links (Di Marzo Serugendo et al. 2010). Self-awareness concept may be related to the recent attempts to create artifacts that have some characteristics typically associated with consciousness (Haikonen 2007, Sanz et al. 2011)

One important differentiation to be made for all the self-properties is the direction.

Self-adaptation and self-management are considered as top-down, whereas self-organization and self-healing are considered as bottom-up properties.

A self-evolvable system will swing between the two directions and benefits from both directions. This kind of swinging was described for architectures shown in Fig. 1.1c.

1.3 Challenges and Limitations

Self-properties represent an important part of complexity engineering. They allow systems to play active and increasingly autonomous roles in accomplishing their tasks, but there are also challenges and limitations to the possibilities of self-properties.

Self-evolvability offers potential solutions to evergrowing complexity but also may be a source of new problems.

Some issues and warnings in implementing self-properties, some solutions and perspectives have been discussed in the literature and will be presented in the following (Herrmann et al. 2005, Frei and Di Marzo Serugendo 2011a, 2011b).

• Inherent Differences between Engineered and Natural Systems

A difference between systems that are engineered and systems that result from natural processes is that engineered systems may be optimized for one or more particular properties with inadequate search for self-evolvability. When the environment in which an engineered system changes or when the uses to which one wants to put an engineered system changes, it is often difficult to change the system to accommodate the new needs.

It was observed that natural systems tend to be much more adaptable than engineered systems (Abbott 2006, 2007). An open problem is to understand the root of this difference and what can be done to surpass such differences.

The increasingly complex systems that self-evolve are not to be re-designed from scratch with every evolvability step. Each step is an evolutionary neighbor of something less complex. The evolutionary sequence provides both a specific feature needed at each step and a framework that can support the evolutionary process itself. The architectures that survive are those that support and facilitate self-evolvable change. Thus the evolutionary process produces not only fitness for changing environments but also itself self-evolvability.

New versions of many engineered systems may be built from scratch, that is, they are not evolved as extensions of previous versions. Those that are extensions of previous versions are often uncoordinated. In the long run, self-evolution rewrites and reduces the structures that impede self-evolution leaving evolvable core frameworks.

It should be emphasized that self-evolvability is not required for any engineering systems. One reason for this is that designing for self-evolvability is very difficult. Another reason is that we do not have a satisfactory way either of specifying or measuring self-evolvability. Given the current state of the knowledge about self-evolvability, the most effective way of requiring self-evolvability is economic. This implies that the developers should evaluate and absorb parts of the cost of post-delivery enhancements.

• Sensitivity to Specific Conditions

Industrial systems often exhibit sensitivity to specific conditions and to disturbances. Certain factors, like energy disruptions or an abnormal increase of temperature and humidity, may lead to system breakdown, while others, as for instance the extreme noise, may have insignificant effects. Some disturbances may have consequences in some cases, lack any effect in others, and may be beneficial in some cases. An automaton using thermal sensors reacts sensitively to changing temperature conditions, whereas an automaton working with optical sensors remains unaffected by temperature.

A degree of randomness may be beneficial for systems structuration in levels.

Systems may efficiently find a way to accomplish their task under certain initial conditions, but may not be able to do so when the conditions are to some extent different. Engineers have to consider their systems' sensitivity to initial conditions and attempt to find solutions to alleviate the effects.

• Time-Scales and Impossibility to Find New Stable States or Convergent Solutions

Most self-evolvable systems can eventually find stable states or stable solutions to achieve their tasks, but this takes time. This means that the designer and the user of self-evolvable systems must be able to accept postponement.

In certain cases, a self-evolvable system may not be able to solve the task given or its calculations may converge too slowly. The engineer has to preview this and arrange for a way out, such as alerting the user and settling for a solution which requires the relaxation of certain constraints or imposing new ones.

The delay in self-evolvability achievement may be due to the inherent difficulty to analyze self-properties. Moreover, the system may find ways to fulfill tasks which the engineer did not plan or preview. The engineer may intend the system to act in a certain way, and in reality, it is all different. Also the interplay between various self-properties may be difficult to analyze and further research efforts may be necessary.

• Reliability and Resilience

Reliability is defined as the probability that an item will perform its intended function for a specified time interval under stated conditions. Reliability analysis addresses the changes in quality of a system over time.

Resilience is the property of a system that enables it to resume its original capabilities after being perturbed. Resilience is a dynamic process that systems exhibit positive behavioral adaptation when they encounter significant sources of stress.

The engineer should be assured that the system does what it is supposed to do, independent from the actual situation and circumstances, and this may be challenging, for some high complexity systems.

Thanks to their redundancy, the systems are often inherently robust to certain types of failures. They may, however, be fragile when facing a different type of fault. An open problem is the lack of widely accepted of what self-evolvability actually requires for reliability and of appropriate standards for unifying the self-evolvability processes.

• Control Issues and Predictability

A common warning that opposes the deployment of complex systems knowledge in engineering is the idea that all control over the system will be lost. However, the issue about self-evolvability is not that the developed systems will be unpredictable, nondeterministic or uncontrolled (Buchli and Santini 2005).

It is expected that the guarantees about the functioning of the system will be of statistical nature. In all engineering works, the guarantees that can be made about a system are limited and essentially of statistical nature. Furthermore, it shows an erroneous understanding of how the development of technology works if a complete understanding of the functioning of a system is demanded before it is accepted. Technology and engineering have worked with systems which are not fully understood but nevertheless they are accommodating and brought good or acceptable services.

Often the understanding is deepened after which it can lead to a yet better or more efficient use. Self-evolvability for a complex system is the way to arrange itself, to find the balance between controllability, predictability, impredicative behavior and a letting loose of some fuzzy aspects of the system (Frei and Di Marzo Serugendo 2011a, 2011b).

Self-evolvability inherently implies loops and circularity. It is known that a graph will contain no cycles or loops if and only if it is well founded (Aczel 1988). This means that a graph describing a complex system that contains loops or cycles is a picture of a non-well-founded set. The presence of cycles and loops would indicate that certain set has itself as a member or that the concept system or definition it models is impredicative.

Impredicative behavior does not mean that the system can't be computable, constructed or engineered (Mossio et al. 2009). For instance, the graphs of a proof may be allowed to contain a balance of cycles and hierarchical trees (Santocanale 2002).

• Relevance Limits

It should be emphasized that self-evolvability is not meant to replace traditional engineering approaches, but is merely a natural development complementing, reactivating and enhancing conventional domains.

Self-evolvability has to be applied to appropriate and critical problems. Even if the engineer will take inspiration and learn from physical and living systems, his main goal is not restricted to understanding the nature. The goal to formulate a model which should serve for understanding the real-world phenomena is the primary concern of the scientist.

Thus if an engineer finds that a few modifications to a system may serve him well, even if these modifications are not justified by complete observations in the real systems, there is no reason why he should not use the modified system and no other justification for the change is needed.

Numerous traditional systems are still operated in a regime where the complexity properties are observed but may be neglected. The reductionistic approach is a powerful method for gaining knowledge about certain aspects of the natural world and for supporting the development of technology precisely because it selects phenomena that have a simple explanation. But the cost of restricting to simplicity only is to be unable to represent the full range of possibilities offered by engineering systems. As a mandatory way of success of thinking and a firm base for undertaking, the reductionism should be complemented by the way of systemic or integrative thinking. The subtle blend of both ways, the reductionistic and the complexity ways, the differential and integrative ways, proved to be the main sources of efficiency.

While, in some cases, traditional engineering studied the ways of taking the complexity out of the systems, we now have to allow complexity to come back in, to complement the reductionism and learn to exploit both epistemological ways for our own good.

Surely, the complexity science and engineering with a self-evolvability perspective in mind will be more goal-directed and successful than conventional reductionistic and distributed attempts (Ottino 2004).

References

Abbott, R.: Complex systems + systems engineering = complex systems engineering. In: Conf. on Systems Engineering Research, California State University, Los Angeles and The Aerospace Corporation. Tech. Rep. (2006)

Abbott, R.: Putting complex systems to work. Complexity 13, 30–49 (2007)

Aczel, P.: Non-well-founded Sets. Stanford, CSLI (1988)

Banzhaf, W.: Self-organizing Systems. Encyclopedia of Physical Science and Technology, pp. 589–598. Academic Press, New York (2002a)

Banzhaf, W.: Artificial Chemistries–Towards constructive dynamical systems. Nonlinear Phenom. Complex Syst. 5, 318–324 (2002b)

Barata, J., Ribeiro, L., Onori, M.: Diagnosis on evolvable production systems. In: IEEE International Symposium on Industrial Electronics (ISIE), Vigo, Spain, pp. 3221–3226 (2007)

Buchli, J., Santini, C.: Complexity engineering, harnessing emergent phenomena as opportunities for engineering. Tech. Rep., Santa Fe Institute Complex Systems Summer School, NM, USA (2005)

Bush, R., Mosteller, F.: Stochastic Models for Learning. Wiley, New York (1995)

De Wolf, T., Holvoet, T.: A taxonomy for self-* properties in decentralized autonomic computing. In: Parashar, M., Hariri, S. (eds.) Autonomic Computing: Concepts, Infrastructure, and Applications, pp. 101–120. CRC Press, Taylor and Francis Group (2007)

Di Marzo Serugendo, G.: Robustness and Dependability of Self-Organizing Systems - A Safety Engineering Perspective. In: Guerraoui, R., Petit, F. (eds.) SSS 2009. LNCS, vol. 5873, pp. 254–268. Springer, Heidelberg (2009)

Di Marzo Serugendo, G., Fitzgerald, J., Romanovsky, A.: Metaself - an architecture and development method for dependable self-* systems. In: Symp.on Applied Computing (SAC), Sion, Switzerland, pp. 457–461 (2010)

Frei, R.: Self-Organization in Evolvable Assembly Systems. PhD. Thesis, Department of Electrical Engineering, Faculty of Science and Technology, Universidade Nova de Lisboa, Portugal (2010)

Frei, R., Di Marzo Serugendo, G.: Advances in Complexity Engineering. Int. J. of Bio-Inspired Computation 1(1), 11–22 (2011a)

Frei, R., Di Marzo Serugendo, G.: Concepts in Complexity Engineering. Int. J. of Bio-Inspired Computation 3(2), 123–139 (2011b)

Haikonen, P.O.: Robot Brains: Circuits and Systems for Conscious Machines. Wiley & Sons, Chichester (2007)

Herrmann, K., Mühl, G., Geihs, K.: Self-Management: The Solution to Complexity or Just Another Problem? IEEE Distributed Systems Online 6(1), 1–17 (2005)

Hofmeyr, J.-H.: The biochemical factory that autonomously fabricates itself: a systems biological view of the living cell. In: Boogard, F., Bruggeman, F., Hofmeyr, J.-H., Westerhoff, H. (eds.) Systems Biology: Philosophical Foundations, pp. 217–242. Elsevier (2007)

Hu, B., Efstathiou, J.: Responsive system based on a reconfigurable structure. In: Pham, D.T., Eldukhri, E.E., Soroka, A.J. (eds.) 2nd I*PROMS Virtual Conference on Intelligent Production Machines and Systems, pp. 517–522 (2007)

Iordache, O.: Polystochastic Models for Complexity. Springer, Heidelberg (2010)

Iosifescu, M., Grigorescu, S.: Dependence with complete connections and applications. Cambridge Univ. Press, Cambridge (1990)

Kelso, J.: The complementary nature of coordination dynamics: Self-organization and the origins of agency. Journal of Nonlinear Phenomena in Complex Systems 5, 364–371 (2002)

Maturana, H., Varela, E.: Autopoiesis and Cognition. The Realization of the Living. Reidel, Dordrecht (1980)

Mossio, M., Longo, G., Stewart, J.: An expression of closure to efficient causation in terms of λ-calculus. J. Theor. Biol. 257, 498–498 (2009)

Mun, J., Ryu, K., Jung, M.: Self-reconfigurable software architecture: design and implementation. In: Proc. 33rd Int. Conf. Computers Industrial Engineering, CIE569, Jeju, Korea, March 25–27, pp.1–6 (2004)

Ottino, J.M.: Engineering Complex Systems. Nature 427, 399 (2004)

Ritchey, T.: Wicked Problems – Social Messes: Decision support Modeling with Morphological Analysis. Springer, Berlin (2011)

Santocanale, L.: A Calculus of Circular Proofs and its Categorical Semantics. In: Nielsen, M., Engberg, U. (eds.) FOSSACS 2002. LNCS, vol. 2303, pp. 357–371. Springer, Heidelberg (2002)

Sanz, R., Hernández, C., Sánchez, G.: Consciousness, meaning and the future phenomenology. In: Machine Consciousness; Self, Integration and Explanation, York, UK, AISB (2011)

Sterritt, R., Hinchey, M.: Autonomic Computing – Panacea or Poppycock? In: The 12th IEEE International Conference and Workshops on the Engineering of Computer-Based Systems, ECBS, pp. 535–539 (2005)

Strunk, J.D., Ganger, G.R.: A human organization analogy for self-* systems. In: Algorithms and Architectures for Self-Managing Systems, San Diego, CA, June 11, pp. 1–6. ACM (2003)

Von Neumann, J.: Theory of Self-Reproducing Automata. University of Illinois Press, Urbana (1966)

Ziegler, G.M.: Lectures on Polytopes. Graduate Texts in Mathematics, vol. 152. Springer, New York (1995)

Chapter 2
General Framework

Abstract. High complexity imposes systems' structuration in multiple levels.

Associating n-categories to levels offers a new perspective for modeling multi-level systems.

The comprehensive framework of polystochastic models, PSM, serving as flexible guideline for self-evolvable systems modeling is introduced.

Polytopic architectures for self-integrative closure and polycategories are presented.

2.1 Categories and Closure

Fundamental studies of closure concern the multi-level structure of reality and its relation to philosophical categories and mathematical categories (Peirce 1956, Hartmann 1952, Poli 2001, Brier 2008, 2009).

Complex systems exhibit hierarchical self-organization in levels under selective constraints. Self-organization will occur when individual independent parts in a complex system interact in a jointly cooperative manner that is also individually appropriate, such as to generate a higher level organization.

Complex systems have a multi-level architecture and this can be observed at different levels of investigation. For example, we can observe an industrial installation at the level of molecules, at the level of devices interactions or as embedded in its environment. The number of observation levels as that of reality levels is finite.

The concept of closure plays a relevant role in biological explanations since it is taken as a naturalized grounding for distinctive biological dimensions, as purposefulness, normativity and functionality. The contemporary application of closure to the biological domain comes from a philosophical tradition tracing back at least to Kant who claimed that living systems should be understood as natural ends, that is as self-organized structures driven by circular and reciprocal causation. The essence of living systems is a form of internal and circular causality between the whole and the parts, distinct from both efficient causality of the physical world and the final causality of artifacts or mechanisms (Kant 1987).

In general systems theory the concept of closure is used to identify or define the system relation with its environment and to explain the autonomy of the systems. Closure and circularity are critical for self-evolvability understanding and managing.

Closure does not mean that the system is not in contact with its environment. Rather the term closure refers to the loop that connects the structures and the functions of individual living-like entities.

It should be emphasized the complementary roles of the closure and opening or disclosure for self-evolvability. The swinging between closure and openness is an important tool for designing creative systems that can autonomously find solutions to highly complex construction problems.

Different closure concepts as organizational closure (Maturana and Varela 1980), closure to efficient cause (Rosen 1991), operational closure (Luhmann 1995), and semantic closure (Pattee 1995, 2001) are important for self-evolvability studies.

According to Pattee, biological organization consists of the integration of two intertwined dimensions, which cannot be understood separately. On the one side, the organization realizes a dynamic and autopoietic network of mechanisms and processes, which defines itself as a topological unit, structurally coupled with the environment. On the other side, it is shaped by the material unfolding of a set of symbolic instructions, stored and transmitted as genetic information.

The dynamic, that is, mechanistic and the informational dimensions realize a distinct form of closure between them, which Pattee labels semantic or semiotic closure. This concept refers to the fact that while symbolic information must be interpreted by the dynamics and mechanisms that it constrains, the mechanisms in charge of the interpretation and the material translation require that information for their own production. Semantic closure, as an interweaving between dynamics and information, constitutes an additional dimension of organizational closure of biological systems, complementary to the operational or efficient one.

In *On a New List of Categories*, Peirce formulates a theory of categories that can demonstrate what the universal conceptions of reality and of thought are.

Peirce's categories are meant to provide a basis for an exploration of a large variety of phenomena, including natural, biological, reasoning and technological.

Peirce proposed an initial list of five philosophical categories: *substance*, *quality*, *relation*, *representation* and *being*.

Later, Peirce discarded substance and being from his initial list of five categories and focused mainly on quality, relation and representation which he called in his technical terms *firstness*, *secondness* and *thirdness*, respectively.

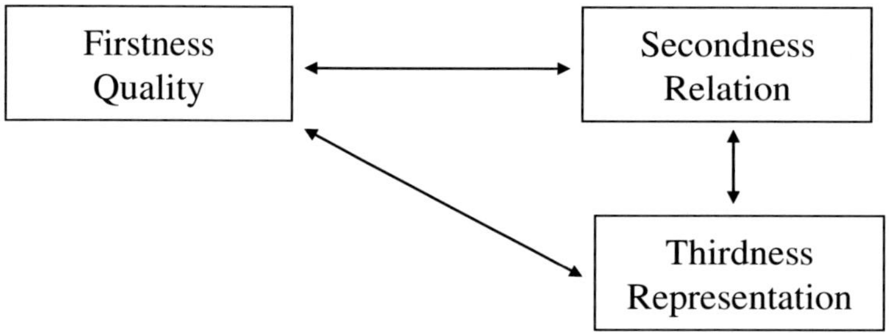

Fig. 2.1 Triadic approach

The triadic architecture of Peirce is illustrated in Fig. 2.1.

Firstness is the conception of existing independent of anything else. Secondness is the conception of existing relative to, the conception of reaction with something else. Thirdness is the conception of mediation, whereby a first and second are brought into relation.

Firstness may be manifested by quality, feeling or multiplicity. Secondness may be manifested by relation, action, reaction, causality, or actuality. Thirdness may be manifested by representation, modality, thought, continuity, unity, or generality.

A triadic approach to biosemiotics has been discussed by Pattee who investigated the physical conditions that are necessary for codes and symbolic controls (Pattee 2001). Pattee introduced the concept of epistemic threshold, the boundary region where local matter has not only its intrinsic physical properties governed by universal laws, but was also about something else. Epistemic matter, in other words, stands for something and the standing for relation is usually considered an emergent process that leads to a triadic Peircean relationship of matter, interpreter and referent.

Studies of emergence, embedding and self-properties for systems of growing complexity suggest reconsidering the Peirce's initial list of five categories.

The categories substance and being have been described by Peirce, as the beginning and end of all conception, respectively. Taking into account the evolution during the years of Peirce's concepts of substance and being a categorical architecture with five levels of reality: substance, firstness, secondness and thirdness, centered by the so-called fourthness identified also as being, or in other context as the Self, was considered (Iordache 2011).

Fig. 2.2 shows a polytopic representation of the Peirce's initial list of five categories and links to the associated Kantian categories, quantity, quality, relation and modality.

This architecture shows being or the Self as a centering category surrounded by substance, firstness, secondness and thirdness. The being considered in its aspect of meta-representation of the four surrounding levels is the key capability for self-evolvability.

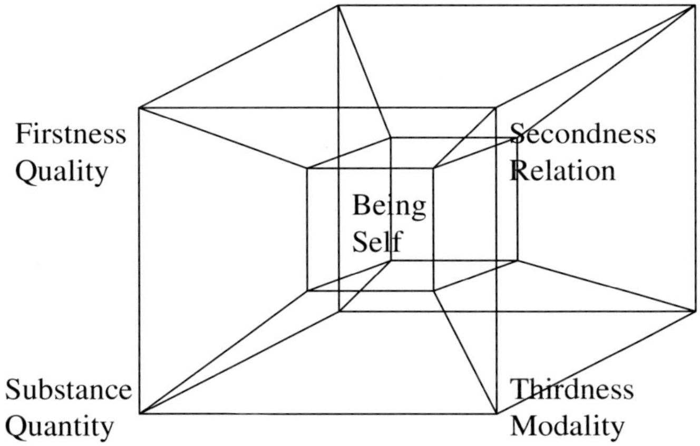

Fig. 2.2 Polytope for five categories

Peirce suggested that cognitive investigation could continue beyond the thirdness. For instance, following the study of existential graphs alpha, α, beta, β, or gamma γ, associated to firstness, secondness and thirdness, Peirce started the study of δ systems, supposed to deal with modals, that is beyond modality (Pietarinen 2003, 2006).

There exist several developments of the Peirce's triadic architectures to tetradic architectures and beyond. An interesting study is the so-called reasoning cycle of Peirce (Sowa 2006). In this case the four modules of the cognitive architecture are: World, Knowledge, Theory and Prediction. The interactions between categories as interpreted by Sowa are induction, abduction, deduction and action. Fig. 2.3 shows a polytope inspired by Peirce's cycle of cognition.

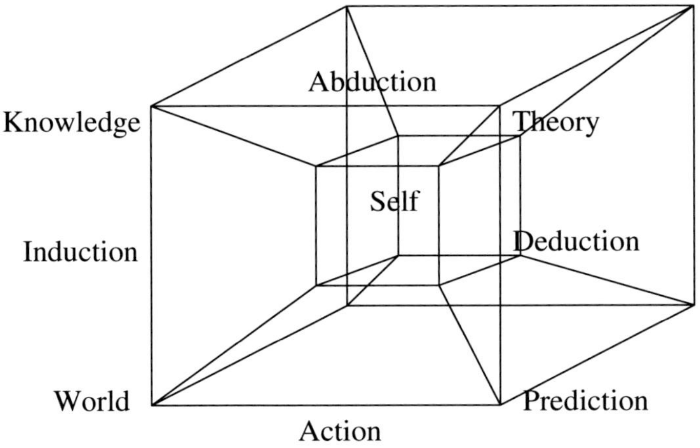

Fig. 2.3 Cognition polytope

Induction or learning starts from observations and looks for commonalities to summarize observed data. Abduction or conjecturing starts with disconnected observations and hypothesizes a theory that relates them. Deduction or inference starts with a theory, observes new data and is used to generate implications.

Staat ascribed the Peirce's categories, firstness, secondness, thirdness to abduction deduction and induction (Staat 1993). Fig. 2.3 shows that abduction is rooted in Knowledge, and deduction is rooted in Theory. If we consider Prediction, and World, together as the category thirdness, the induction will appear to be rooted here.

Taking inspiration from Peirce's philosophy, Brier formulated a transdisciplinary theory of information, semiotics, consciousness and cultural social communication illustrated by the fourfold cybersemiotic star (Brier 2008). Fig. 2.4 shows a polytope based on the Brier's cybersemiotic star.

The four legs correspond to the four main areas of knowledge that is: Material, Living, Consciousness and Mentality. A comparison with the Hartmann's ontological hierarchy is of interest (Hartmann 1952).

The center of the cybersemiotic star was related by Brier to semiotic mind. It may be considered as a meta-representation of the fourfold star and it is linked to self-evolvability.

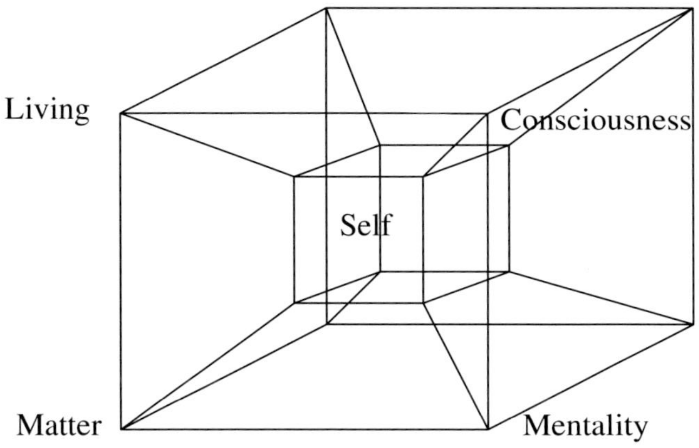

Fig. 2.4 Cybersemiotic polytope

This central category, the Self, is crucial for the transition from evolvable systems to self-evolvable ones. Briers' cybersemiotic star may be interpreted as a 2D-projection of a more general cybersemiotic polytope.

2.2 General PSM Framework

Polystochastic models, PSM, study started by considering complex systems to be compound processes organized hierarchically in levels as arrays of systems within systems (Iordache 1987).

The PSM is developed now as a modeling tool for high-level complexity, mainly for evolvable and self-evolvable systems investigation. The complexity was portrayed in PSM studies using concepts such as hierarchy and conditioning levels, real and formal or in other words non-standard time and probability algebraic frames, and by methods as categorification and integrative closure. Conventional methods, applied in specific ways, joined new ones resulting in a distinctive understanding of complexity (Iordache 2010).

The elements of basic PSM frame are quadruple of vectors [S, K, U, P] denoted also SKUP. The notations are: S-States, K-Conditions, U-Operators, and P-Possibilities.

Observe that the early SKUP framework involves only two levels or realms, S and K. The relation with random systems with complete connections theory is clearly identifiable (Iosifescu and Grigorescu 1990). But it should be emphasized that the two level architectures have limited efficiency for high complexity problems.

As for other approaches to complexity, it was assumed that the complexity can be managed through supplementary hierarchical layering. The elements of SKUP have been considered as vectors. The conditioning levels have been correlated to time and to space scales (Iordache 2011).

Each component of the vectors corresponds to a different conditioning level and a different time scale.

The basic elements of the SKUP have been considered as vectors:

$$S = (s^0, s^1, \ldots, s^n, \ldots, s^M);\ K = (k^0, k^1, \ldots, k^n, \ldots, k^M);$$
$$U = (u^0, u^1, \ldots, u^n, \ldots, u^M);\ P = (p^0, p^1, \ldots, p^n, \ldots, p^M).$$

Here s^n represents the particular state at the level n, and k^n represents the particular condition at the level $n \leq M$. Upper indices are reserved to levels, while lower indices are reserved to time steps. The components of U are operators such as: u^n: $k^n \times s^{n'} \rightarrow s^{n''}$

PSM should describe parallel evolutions. Moreover, S and K are associated to different types of algebraic fields. Despite algebraic framework differences, S and K are interconnected. This interconnection is described by operators U and possibilities P.

U characterizes the K to S transition and P characterizes the S to K transitions, that is:

$U: K \rightarrow S$ and $P: S \rightarrow K$.

Operators U should be able to describe change of conditioning level and splitting of levels. Possibilities P replacing and generalizing probabilities have been studied in game theory (Hammond 1994), in fuzzy logic (Dubois and Prade 2001), and in other domains.

The possibilities P may be defined by vectors such as:

$$P(K) = (p(k^0), p(k^1), \ldots, p(k^m), \ldots, p(k^M)).$$

The component $p(k^n)$ is an evaluation of the condition k^n.

An innovative aspect for PSM concerns the differential model for K defined process. The elements of K are resulting as solutions of differential equations (Iordache 2009, 2010).

These models have been used as generic models producing other models.

The last development stage for PSM, the categorical approach, appears as a categorification of stochastic transition systems for growing n-dimensional problems.

Categories are linked to the different levels of reality. The notion of level or reality which was firstly studied from an intuitive point of view may be approached from a more formal point of view based on category theory. In this case, the levels and sub-levels of reality are characterized and distinguished by their categories and sub-categories.

A category is specified by objects and arrows called also morphisms.

In numerous situations the physical systems are objects while the morphisms correspond to processes. For PSM frameworks the conditions K may represent the category describing the types of component processes. The process types are the objects of category. Interactions among types can be modeled as morphisms.

The arrows, that is the morphisms, describe the transition relations between the states of the component processes. Different algebraic frameworks for states-S (dynamical, analogical, and physical) and conditions-K (symbolic, digital, and formal) have been considered.

One can define a special category which has categories as objects. The morphism between two categories is then defined as functor. Functors as denoted by U are accounting for interactions in K, and between categories K and S. Other functors as the possibilities P supplement the probabilities to express potentiality, fuzziness, uncertainty, and emergence.

Advancements in modeling higher complexity, the evolvability request, required to take into account multiple levels and multiple SKUPs interaction.

Replacing K by several levels is mandatory for the study of higher complexity.

Instead of the category K, the categories K1, K2 and K3 have been considered as a preliminary development of SKUP (Iordache 2010).

Centered, four realms PSM frameworks, resulting by integrative closure, have been presented as the architecture shared by numerous autonomous systems (Iordache 2011).

The pursuit of self-evolvability for systems imposes the polytopic architecture associated to the general PSM framework as shown in Fig. 2.5.

The basic levels and categories are S, K1, K2 and K3 represented on the front face of the outer cube and S′, K1′, K2′ and K3′ represented on the back face of the outer cube.

The swinging between the two faces of the outer cube is mediated by the inner cube identified as the Self.

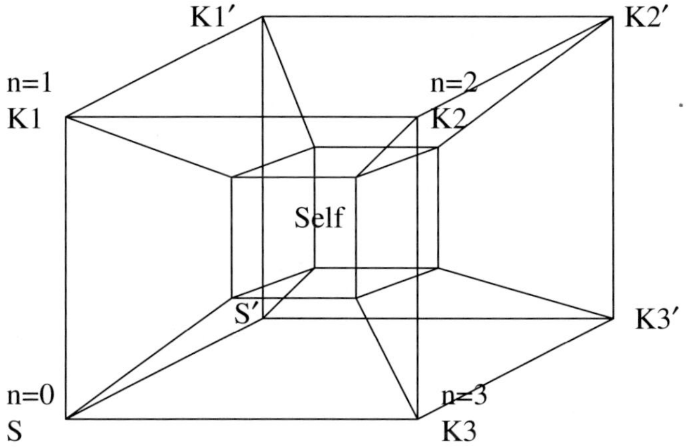

Fig. 2.5 Polytope for general PSM framework

Notable theoretical perspectives resembling the PSM approach are the memory evolutive systems (Ehresmann and Vanbremeersch 2007) and the hierarchical cycles (Louie and Poli 2011). These research directions rely on the power of category theory, too. They develop, in different ways, the idea of iterative constructions of systems over systems in which the system of different layers presents specific properties.

2.3 Self-Integrative Closure

Integrative closure appeared as the direct consequence of mutual restrictedness or exclusiveness of the new levels relative to the previous ones, and of the finite number of levels to be considered. Integrative closure approach is not looking for an identity between the philosophical and mathematical categorical viewpoints but for a structural analogy and a common methodology shared by different domains as knowledge organization, problem solving or technological developments (Iordache 2010).

The conventional hierarchical structures cannot serve as general models for multiple-level knowledge organization. Confronting higher complexity the task of knowledge integration remains pertinent.

The hierarchical structure should be closed and replaced by a network. Finally, it is generally acknowledged that both trees and cycles are necessary. It is the timing of activation and blend of both that matters.

Fig. 2.6 shows the polytope of categories and illustrates self-integrative closure concept.

Fig. 2.6 proposes an extended structural analogy that of the hypothetical integrative closure architecture including philosophical categories architectures as studied by Peirce (substance, firstness, secondness, thirdness and fourthness) and the mathematical n-categories (n=0, 1, 2, 3 and 4). Fig. 2.6 outlines the links to Kantian categories, quantity, quality, relation and modality.

We refer to this overarching framework as *self-integrative closure*.

Adoption of the n-categorical standpoint, suggested the initial extending the investigation to four levels or realms. Support for the four-level architectures is offered by different domains. A primary source is in data processing and neurodynamics (Cowan 2000). According to Cowan the capacity of short-term memory is limited to the number of four items to which attention can be simultaneously directed. There exists a central capacity limit of four chunks in short-term explicit memory presumably corresponding to the focus of attention. This theory assumes that attention is depending on oscillation of cortical potentials. A cortical wave of about 10 Hz is supposed to select items from a large short-term store. Other wavelets at a frequency at about 40 Hz then select one item each. Such considerations do not exclude to direct attention to more than four items or realms, but the resulting processes may be transient.

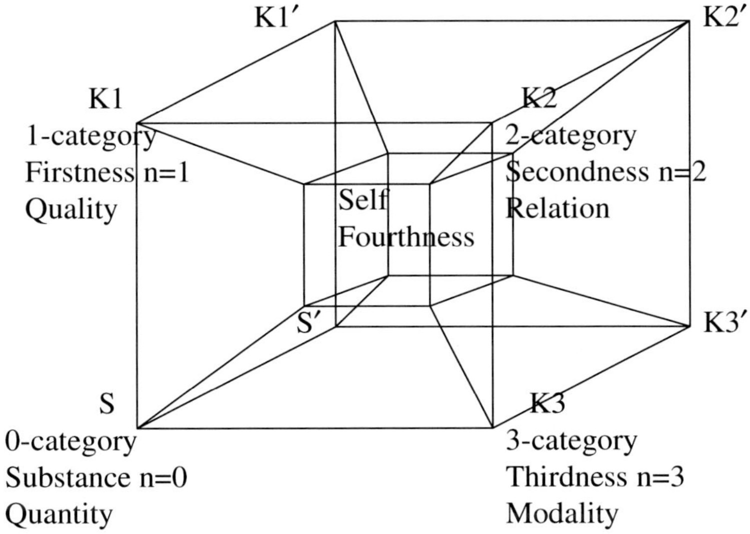

Fig. 2.6 Polytope for self-integrative closure

Significant support for the four levels or four categories architectures in data processing is given by mathematical category theory too (Leinster 2004). Apparently complexity of n-categories rises with n, dramatically. Developing the study of centered four levels means to include the categorical approach to 4-categories.

The four levels are associated in increasing order of complexity starting from 0-category that is from sets, to 1-category that is to conventional categories, to 2-categories, and then to 3-categories. The internal cube, the Self, is associated in this representation to n-categories with $n \geq 4$.

The difficulty to work with mathematical n-categories is that as the number of dimensions increases the complexity of the necessary rules to be specified increases rapidly. For one dimension the rules may be written down on one line, and those for two dimensions may be expressed in diagrams occupying a typical page. For four dimensions the detailed diagrams are so large that they will not fit in acceptable sized book. The 4-category diagram techniques just start to be developed. The difficulty of presentation was considered as a supplementary reason to restrict the majority of studies to 3-categories (Iordache 2010). However we cannot exclude 4-categories or higher ones in the long run. Clearly some other ways of approaching and presenting the theory should be envisaged.

The *self-integrative closure* is based on the hypothesis that there exists a structural correlation between philosophical and mathematical categorification architectures.

As shown in Fig. 2.6, the four levels of reality or the four philosophical categories of Kant or Peirce have been associated to the corresponding mathematical n-categories.

Philosophical categorification is the philosophical counterpart of categorification introduced in mathematics, but replacing logical concepts for categorical concepts, and also set-theoretic notions by category-theoretic notions in order to investigate concepts. The categories are attempts to distill the essence of a certain domain. This is also the goal of people working in that domain.

Category theory could serve as a *lingua franca* that lets us translate between certain aspects in different domains and eventually build a general science for complex systems and processes (Baez and Stay 2008).

Traditionally philosophical categories were not studied in terms of mathematical n-category theory.

The significance of the hypothetical structural analogy between categorical approach in philosophy and mathematics needs more study. The fundamental problem of categorification and decategorification was discussed by Kant (Kant 1987). Kant distinguished two ways of analysis, a qualitative one and a quantitative one. The first may be linked to the relation between conditioned and the condition that is, to the way S→K1→K2→K3 while the second corresponds to the way K3′→ K2′→K1′→S′, from the whole to the parts (Fig. 2.6).

The need for both epistemological ways finds a strong support in the studies of metastable coordination dynamics of the brain (Kelso 2002, Kelso and Tognoli 2009). Metastability has been highlighted as a new principle of behavioral and brain function and may point the way to a truly complementary neuroscience. From elementary coordination dynamics it was shown explicitly that metastability is a result of a symmetry-breaking caused by the subtle interplay of two forces: the tendency of the components to couple together and the tendency of the components to express their intrinsic independent behavior. The metastable regime reconciles the well-known tendencies of specialized brain regions to express their autonomy, that is differentiation, and the tendencies for those regions to work together as a synergy, that is integration.

Nevertheless, the architectural similarities between philosophical categories and mathematical category theory cannot be interpreted as a coincidence. Peirce, inspired by Kant, is acknowledged today as a precursor of higher-dimensional algebra and in this way of n-category study. So, it would be interesting to re-evaluate Peirce's work about categories, in terms of mathematical n-categories. This will relate n-categories to pragmatism as founded by Peirce. What may be called categorical pragmatism refers to Peirce's fundamental concern to discover the basic elements or principles essential in the process of inquiry, rather than to just formulate a criterion of truth by means of which the results of inquiry are to be judged for their truth value.

Table 2.1 summarizes the categorification aspects for PSM frameworks.

The study of PSM framework for self-integrative closure and the emergence of the Self corresponding to $n \geq 4$ represent a challenge from both conceptual and mathematical points of view.

Table 2.1 Categorification for PSM framework

Level	S (K0)	K1	K2	K3	Self
-	n=0	n=1	n=2	n=3	n=4
Categories	0-category	1-category	2-category	3-category	4-category
Example	sets	Set	Cat	Fun	-

For *n*-category theory, a category such as Set is a 1-category, with 0-objects that is sets, for objects and 1-morphisms, that is functions, for arrows (Appendix 1). A functor is the morphism between categories. Actually a functor between two categories is also defining as mapping objects and morphisms of one category to objects and morphisms of the other, in such a way that the morphism between two objects is mapped to the morphism between two mapped objects. Thus a functor appears as a transformation which keeps the basic structure. The category of categories, Cat, has categories for objects and functors for arrows. Thus, a functor is a 2-morphism between 1-objects, that is 1-categories, in a 2-category.

One can define a new category with functors as objects. A natural transformation is the morphism between functors. The general idea is to transform not only the underlying categories one into another but also to parameterize that transformation by any basic constituting element, translating the idea that a global transformation between complex architectures is made by local transformation with different levels of accuracy.

The functor category, Fun, has functors as objects and natural transformations as arrows. Thus, a natural transformation is a 3-morphism between 2-objects, that is functors, in a 3-category.

Using n-category theory we propose a unified framework that allows describing in a condensed manner the transformations allowed in the complex system.

The categorical approach highlights the possible transformations, given a structure of the system, and checks formally the analogy or the similarity between architectural organizations.

2.4 PSM and Polycategories

Since PSM aim to describe the composition of several processes they may be presented in the frame of polycategories (Appendix 4).

Polycategories were introduced by Lambek and Szabo with the intention of providing a categorical framework for classical logic, with multiple formulae on both left and rightsides of the sequent (Lambek 1969, Szabo 1975). Their composition law is based on the cut rule of logic.

A category allows having morphisms which go from single objects to single objects.

A polycategory allows having morphisms from lists of objects to lists of objects.

A typical morphism in a polycategory called also polymorphism or polymap would be denoted: $f: X_1, X_2, \ldots, X_n \rightarrow Y_1, Y_2, \ldots, Y_m$

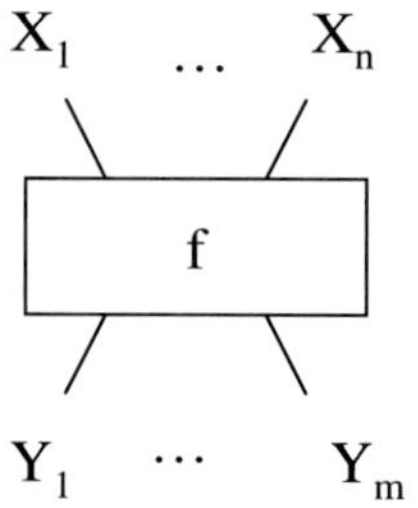

Fig. 2.7 Polymap

Fig. 2.7 shows a polymap. Here f denotes a process. The domain channels are on the top and the codomain channels on the bottom. A process acts on a number of channels by either accepting input events or producing output events in accordance with the rules or protocols associated to each channel.

Physically one could think of the channels as pipes for chemical engineering installations or wires for electrical circuits and so on.

If a codomain channel α of a process f, and a domain β of another process g, share a common protocol then f and g may be composed on α and β to form a new process.

Fig. 2.8 shows a composed process. The double lines conventionally represent strings of channels.

Such interpretations highlight the close relation between polycategories and polystochastic models, PSM, introduced as composed processes (Iordache 1987).

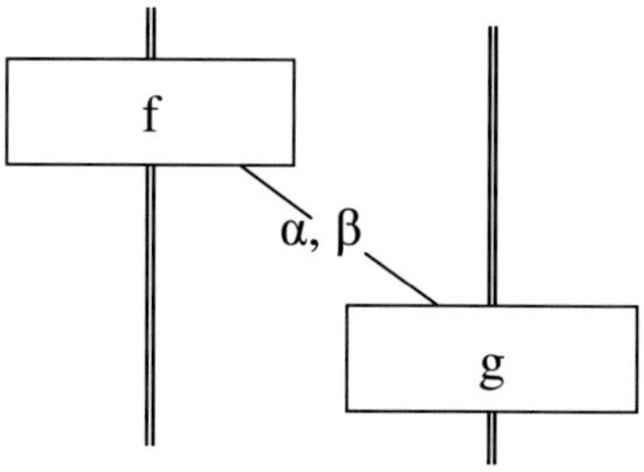

Fig. 2.8 Composed processes

There are several contexts in which the generalization related to the concept of polycategories would be useful.

As a first example consider vector spaces or any class of modules in which one can form a tensor product. A polycategory, having as objects such spaces, can be defined. The morphism of the above form may be a linear function:

$$f: X_1 \otimes X_2 \otimes \ldots \otimes X_n \rightarrow Y_1 \otimes Y_2 \otimes \ldots \otimes Y_m \tag{2.1}$$

Such polycategories have proven to be useful in the analysis of ordinary categories in which one can form tensor products of objects. Categories in which one has a suitable notion of tensor product are called monoidal.

Another significant application of polycategories is to logic. The interest is in the analysis of sequents, written: $X_1, X_2, \ldots, X_n \vdash Y_1, Y_2, \ldots, Y_m$

Here $\vdash$ denotes the implication and $X_1, X_2, \ldots, X_n, Y_1, Y_2, \ldots, Y_m$ may represent formulas in some logical system.

The above sequent holds if and only if the conjunction of $X_1, X_2, \ldots, X_n$ logically entails or implies the disjunction of $Y_1, Y_2, \ldots, Y_m$.

There is a correspondence between the sort of logical entailments considered here and categorical structures (Lambek and Scott 1986).

Notice the difference between this and the first example. When considering vector spaces, the commas on the left and rightsides were both interpreted as the tensor product. However for the logically inspired example, there are two different interpretations. Commas on the left are treated as conjunction, while commas on the right are treated as disjunction.

Thus for a categorical interpretation of polycategories one needs categories with two monoidal structures that interact in an appropriate fashion. Such categories are called linearly or weakly distributive (Cockett and Seely 1997) and they represent the appropriate framework for linear logic (Girard 1987)

Suppose we are given two polymorphisms of the following form:

$$f: X_1, X_2, \ldots, X_n \rightarrow Y_1, Y_2, \ldots, Y_m, C \quad (2.2)$$

$$g: C, Z_1, Z_2, \ldots, Z_k \rightarrow V_1, V_2, \ldots, V_j, C \quad (2.3)$$

The polymorphisms f and g are considered as processes in the PSM terminology.

Note the single object C common to the codomain of the process f and the domain of the process g. Then under the definition of polycategory, we can compose these to get a morphism of form:

$$g \circ f = X_1, X_2, \ldots, X_n, Z_1, Z_2, \ldots, Z_k \rightarrow Y_1, Y_2, \ldots, Y_m, V_1, V_2, \ldots, V_j \quad (2.4)$$

The object C which is eliminated after composition is called the cut object, a terminology derived from logic. The morphism g o f is a compound process. C controls this composition.

Fig. 2.9 illustrates the composition in polycategories.

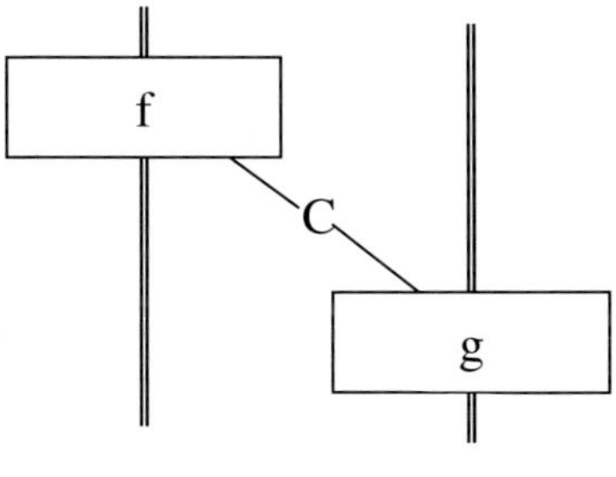

Fig. 2.9 Composition for polycategories

Composition is represented by the concatenation of the processes f and g followed by joining the incoming and outgoing edges corresponding to the cut object C.

References

Baez, J., Stay, M.: Physics, Topology, Logic and Computation: A Rosetta Stone. In: Coecke, B. (ed.) New Structure for Physics. Lecture Notes in Physics. Springer (2008)

Brier, S.: Cybersemiotics: Why information is not enough. University of Toronto Press, Toronto (2008)

Brier, S.: Cybersemiotic pragmaticism and constructivism. Constructivist Foundations 5(1), 19–39 (2009)

Cockett, J.R.B., Seely, R.A.G.: Weakly distributive categories. J. Pure Appl. Algebra 114, 133–173 (1997)

Cowan, N.: The magical number 4 in short-term memory: A reconsideration of mental storage capacity. Behavioral and Brain Sciences 24, 87–185 (2000)

Dubois, D., Prade, H.: Possibility theory, probability theory and multiple valued logics: a clarification. Annals of Mathematics and Artificial Intelligence 32, 35–66 (2001)

Ehresmann, A.C., Vanbremeersch, J.-P.: Memory evolutive systems: hierarchy emergence, cognition. Elsevier, Amsterdam (2007)

Girard, J.-Y.: Linear logic. Theoretical Computer Science 50, 1–102 (1987)

Hammond, P.J.: Elementary Non-Archimedean Representation of Probability for Decision Theory and Games. In: Humphreys, P. (ed.) Patrick Suppes: Scientific Philosopher. Probability and Probabilistic Causality, vol. I, pp. 25–59. Kluwer Academic Publishers, Dordrecht (1994)

Hartmann, N.: The new ways of ontology. Greenwood Press, Westport (1952)

Iordache, O.: Polystochastic Models in Chemical Engineering. VNU-Science Press, Utrecht (1987)

Iordache, O.: Evolvable Designs of Experiments Applications for Circuits. J Wiley VCH, Weinheim (2009)

Iordache, O.: Polystochastic Models for Complexity. Springer, Heidelberg (2010)

Iordache, O.: Modeling Multi-Level Systems. Springer, Heidelberg (2011)

Iosifescu, M., Grigorescu, S.: Dependence with complete connections and applications. Cambridge Univ. Press, Cambridge (1990)

Kant, I.: Critique of Judgment. Hackett Publishing, Indianapolis (1987)

Kelso, J.: The complementary nature of coordination dynamics: Self-organization and the origins of agency. Journal of Nonlinear Phenomena in Complex Systems 5, 364–371 (2002)

Kelso, J.A.S., Tognoli, E.: Toward a Complementary Neuroscience: Metastable Coordination Dynamics of the Brain. In: Murphy, N., Ellis, G.F.R., O'Connor, T. (eds.) Downward Causation and the Neurobiology of Free Will. UCS, vol. 3, pp. 103–124. Springer, Heidelberg (2009)

Lambek, J.: Deductive systems and categories (ii). In: Hilton, R.J. (ed.) Category Theory, Homology Theory and their Applications, I. Lecture Notes in Mathematics, vol. 87, pp. 76–122. Springer (1969)

Lambek, J., Scott, P.: Introduction to Higher-order Categorical Logic. Cambridge Univ. Press (1986)

Leinster, T.: Higher Operads, Higher Categories. Cambridge University Press (2004)

Louie, A.H., Poli, R.: The spread of hierarchical cycles. International Journal of General Systems 40(3), 237–261 (2011)

Luhmann, N.: Social Systems. Stanford University Press, California (1995)

Maturana, H., Varela, E.: Autopoiesis and Cognition. The Realization of the Living. Reidel, Dordrecht (1980)

Pattee, H.H.: Evolving self-reference: matter, symbols, and semantic closure. Communication and Cognition –Artificial Intelligence 12(1-2), 9–28 (1995)

Pattee, H.H.: The Physics of Symbols: Bridging the Epistemic Cut. Biosystems 60, 5–21 (Special Issue: Reflections on the work of Howard Pattee) (2001)

Peirce, C.S.: Collected Papers of Charles Sanders Peirce, vol. 1-8. Cambridge University Press, Harvard (1956)

Pietarinen, A.-V.: Peirce's magic lantern: moving pictures of thought. In: 2003 Conference of the American Philosophical Association, Eastern Division, Annual Meeting of the Charles S. Peirce Society, Washington (December 2003)

Pietarinen, A.-V.: Signs of Logic. Peircean Themes on the Philosophy of Language, Games and Communication. Springer (2006)

Poli, R.: The basic problem of the theory of levels of reality. Axiomathes 12, 261–283 (2001)

Rosen, R.: Life itself: A comprehensive enquiry into the nature, origin and fabrication of life. Columbia University Press, New York (1991)

Sowa, J.F.: The Challenge of Knowledge Soup. In: Ramadas, J., Chunawala, S. (eds.) Research Trends in Science, Technology and Mathematics Education, pp. 55–90. Homi Bhabha Centre, Mumbai (2006)

Staat, W.: On abduction, deduction, induction and the categories. Transactions of the Charles S. Peirce Society 29, 225–237 (1993)

Szabo, M.E.: Polycategories. Comm. Alg. 3, 663–689 (1975)

Chapter 3
Differential Models

Abstract. The models presented capture the intuition of the elementary differential calculus and provide the theoretical substrate for studying designs, schemas, shapes and constructions.

Unconventional frames for time and space, as for instance Galois fields or cyclic groups, have been used to describe the finite or cyclic type of separation and classification processes.

A wave equation is proposed as a differential model for separation and pattern recognition. This model is an abstract complement of transfer equations. The model generates design of experiment matrices as solutions.

Differential posets are introduced as powerful tools in the study of high complexity.

The connection with dual algebras is emphasized.

The notion of a differential category provides a basic axiomatization of differential operators for categories.

3.1 Cyclic Framework

The adopted point of view is that the significance of the process parameters should agree firstly with the mechanism, the nature and goals of analysis for the studied system.

Less-conventional mathematical frameworks are acceptable if these frames can naturally describe the system evolution and the system analysis can proceed on this basis.

Evolvable systems study needs appropriate concepts for time and space (Iordache 2009, 2010, 2011).

Specific chronotopoids should be associated to different reality levels (Poli 2007).

Different kinds of time and of space should be considered in multi-level modelling.

Unconventional models for space of conditions, K, will be presented in the following.

Algebraic finite frames, such as Galois fields and rings or cyclic groups, represent mathematical frameworks that have been used to describe the finite, logical or cyclic type of cognition processes.

A cyclic framework complementing the usual linear one from classical physics proves to be necessary. Evolvability description requires the slow time or cyclic conditions K-processes and the faster, dynamical or linear states S-processes.

Algebraic finite fields represent a common choice for conditions K, whereas the real field is the commonplace structure for states, S. There is a natural hierarchical or cyclic structure associated to finite fields and this explains why they are considered as the appropriate tool for systems structured in conditional levels.

An equation describing the cognitive and self-evolvable systems would contain parameter analogues to the space and the time from the dynamical mathematical models known from physics and engineering.

Consider for example the space Z of vectors describing the properties of an object to be classified and the time T describing the degree of advancement of the pattern recognition, classification, or development, for that object. For the classification process it is possible to associate to different steps in a classification schema digits such as "0" or "1", with significance as no or yes, true or false, separated or non-separated, identified or non-identified (Iordache et al. 1993a, 1993b, 1993c).

To any object to be classified, a vector Z will be associated in which the properties are specified by digits in the hierarchical order of significance for classification. The same vector Z will give a description of the classification stages in the associated pattern recognition schema. Z describes pattern recognition or stages in problem-solving or development stages for systems and so forth. Denote $Z=z_0z_1\ldots z_j$.

The component z_j should specify the presence of an attribute in classification step, its absence but also partial or uncertain results.

The mathematical framework for Z cannot be limited to that of dyadic that is to Boolean calculus. The need for multi-valued characterization of classification steps and of objects, the uncertainty, imposes mathematical tools completing the better studied dyadic calculations. Detailed description of dynamical systems needs vector characterization corresponding to multi-valued logic such as: "0", "1", "2" and so on, meaning, for instance, low, average, high and so on. The coordinate "z_j" characterizes the properties and also it is naturally associated to a stage of classification schemas that make use of the difference in properties noted by "z_j" to perform that kind of classification, pattern recognition, or development.

The degree of advancing in the classification, pattern recognition or in development, denoted by T, was defined as the necessary level of similarity T, between two objects representation, to be classified in the same class (Iordache et al. 1993 a, 1993c). It may be an expansion of the type: $T = t_0t_1\ldots t_j$ with the digits $t_j = 0, 1, 2$ and so on. Denote also this vector by $T= (t_j)$. Each value of T corresponds to another potential step in pattern recognition or in development. Single component vectors with modulo-m algebra structure will be presented as a first example. This is one of the weakest algebraic structures for T and Z still providing a mathematically tractable model adequate to classification and pattern recognition operations or to development study. A slightly different framework to be considered is that of Galois finite fields. Recall that finite fields with the same number of elements are isomorphic.

Examples of addition and product tables are presented in Table 3.1 and Table 3.2. Here "⊕" denotes the addition and "⊗" denotes the product. The sum and product refers to component-wise operations for vectors as Z or T in K. Here C (m) denotes the modulo-m algebraic framework and GF (m) the Galois field of order m.

C (m) enables to fit the physical intuition concerning the cyclic character of the classification operations in m steps and to justify this first choice for algebraic framework. If m=2 the sum ⊕ is defined as follows: for any two elements T= (t_j) and S= (s_j) the dyadic sum is: $(t \oplus s)_j=((t_j+ s_j) \bmod 2)$. This means that 1⊕1=0, 1⊕0=1. The sum is the dyadic addition, ⊕ equivalent to the dyadic difference. The rule of addition ⊕ signifies that two identical digits have no effects for classification. Only the difference in digits makes a contribution. This addition appears rather as a comparison than as a sum. The product ⊗ is introduced in a way related to cyclic operations too. Product definition takes into account that after m steps the classification process restarts. For the time T or space Z, no change should happen after completion of a cycle of classification operations.

An elementary algebraic framework to be considered is the Galois finite field, GF (m). If m is not a prime number we are faced with rings instead of fields.

Table 3.1 Sum and product in C (m)

C (2)

(x+y) mod2			(x.y) mod2		
⊕	0	1	⊗	0	1
0	0	1	0	0	0
1	1	0	1	0	1

C (3)

(x+y) mod3				(x.y) mod3			
⊕	0	1	2	⊗	0	1	2
0	0	1	2	0	0	0	0
1	1	2	0	1	0	1	2
2	2	0	1	2	0	2	1

C (4)

(x+y)mod4					(x.y)mod4				
⊕	0	1	2	3	⊗	0	1	2	3
0	0	1	2	3	0	0	0	0	0
1	1	2	3	0	1	0	1	2	3
2	2	3	0	1	2	0	2	0	2
3	3	0	1	2	3	0	3	2	1

This algebraic framework was extensively applied in formal logics. For illustration purposes, the operations in GF (3) and GF (4) are presented in Table 3.2.

Let Y denotes the range, the output of a system that performs classification based on features or property examination. Y is element of the same algebraic frames as T or Z. Y may be single-dimensional vector and may assume values 0, 1, 2 and so on, corresponding to various outputs. Multi-dimensional values like $Y=y_0y_1y_2\ldots y_j$ should be examined too. Y, as T or Z, is represented by finite strings. Y definition needs to ensure the logical consistency of the framework.

Appropriate algebraic structures for the range of Y are algebras or fields such as the field of real numbers, the modulo-m algebras, or the finite Galois field, GF (m) that provides physically significant and mathematically tractable models.

Table 3.2 Sum and product in GF (m)

GF (2)

(x⊕y)			(x⊗y)		
⊕	0	1	⊗	0	1
0	0	1	0	0	0
1	1	0	1	0	1

GF (3)

(x⊕y)				(x⊗y)			
⊕	0	1	2	⊗	0	1	2
0	0	1	2	0	0	0	0
1	1	2	0	1	0	1	2
2	2	0	1	2	0	2	1

GF (4)

(x⊕y)					(x⊗y)				
⊕	0	1	2	3	⊗	0	1	2	3
0	0	1	2	3	0	0	0	0	0
1	1	0	3	2	1	0	1	2	3
2	2	3	0	1	2	0	2	3	1
3	3	2	1	0	3	0	3	1	2

The dyadic differential calculus was initially developed for GF (2) situations (Harmuth 1977). If m is a prime-p, the range Y is the standard framework for multi-valued logic. Single-dimensional vectors T, Z, Y are useful if the classification process is based on a single property. For multiple-level cases the parameters T, Z and the functions as Y will be tensor products of single-level cyclic groups.

3.2 First Order Wave Equation

The basic material balance for chemical engineering transport processes is represented by the first-order wave equation:

$$\frac{\partial y}{\partial t} + v\frac{\partial y}{\partial z} + q(y) = 0 \qquad (3.1)$$

Here y (t, z) may denote a concentration variable in t-time and z-space, v denotes the velocity, and q(y) denotes a separation rate (Rhee et al. 1989).

The model (3.1) describes many phenomena of physical and technological interest, as momentum, heat and mass transfer.

For mass transfer, the basic model (3.1) shows that the variation of concentration in time is due to the convective process with velocity v and to the kinetic process of interaction, q(y). All the parameters of the model (3.1) are defined on the mathematical real field.

For physical systems study, the real field plays the dominant role.

However this field may be unsuitable to describe some features of high complexity systems.

Paralleling (3.1), a differential model for the space of conditions K, results.

We start with a process of classification, where K is the space of classes.

Classification and the judgment of similarity are fundamental in cognition, serving as the basis for actions. The classification, separation and pattern recognition are the key ingredients in data processing and in problem solving for both natural and artificial evolvable systems. Living or artificially living systems do not survive in environments that they do not recognize or misclassify. Living supposes identification, classification or categorization, separation or combination.

Preliminary attempts for classification or pattern recognition modeling by differential equations outlined the major role of orthogonal arrays (Iordache 2009). A significant result was that the pattern recognition methods parallel screening procedures in experiment design and in problem solving. In particular cases one obtained as solutions of the first-order wave equation orthogonal arrays matrices, Walsh-Hadamard matrices, or Latin squares. Models of cognitive processes such as pattern recognition prove to have as solutions logical thinking methods as that applied in designs of experiments. The result emphasized the deep relation between cognition and evolvability as presented in constructivist perspective and the assertion that both cognition and evolution should be based on similar sets of techniques and models.

To establish the analogs of a dynamical model for classification or pattern recognition, the concept of time and of space in the conditions space K, was adapted to the objectives of the present study.

One of the simplest mechanisms of pattern recognition, classification or development is that in which small changes of the degree of pattern recognition, ∂T, are associated to small changes of the answer, ∂Y. It should be emphasized that the differential is in fact a difference since T and Y are discrete.

Moreover, the change of answer ∂Y depends on both the existing answer Y and the change ∂T of T that is:

$$\frac{\partial Y}{\partial T} \propto Q(Y) \tag{3.2}$$

It is supposed that ∂T is non-null. Otherwise the differential equations are replaced by difference equations. The rate of pattern recognition or classification is denoted by Q. This mechanism is of kinetic type.

Another classification mechanism takes into account that the variation of the answer Y, along the degree of recognition T, is proportional to the answer variation along the features space Z. Classification, pattern recognition and development mean in fact travel in time T, along the space of properties, Z. As Z is screened with a velocity V, the degree of pattern recognition varies proportionally.

This means that:

$$\frac{\partial Y}{\partial T} \propto V \otimes \frac{\partial Y}{\partial Z} \tag{3.3}$$

Here the velocity is a vector $V = v_0 v_1 v_2 \ldots v_j$ or $V = (v_j)$. This mechanism is of convection or drift type.

The general model of the pattern recognition process including both types of recognition processes, corresponding to the evolution according to T and Z, is the first order wave equation WE:

$$\frac{\partial Y}{\partial T} \oplus V \otimes \frac{\partial Y}{\partial Z} \oplus Q(Y) = 0 \tag{3.4}$$

The initial condition is:

$$Y(Z, 0) = F(Z) \tag{3.5}$$

Obviously V and Q may depend on T and Z.

The fact that the addition is equivalent to the difference suggests that a second-order wave equation does not give new solutions in K, as defined.

The mathematical formalism for modeling conditions K apparently follows that of the real states S as a first-order wave equation, WE, but with different addition and product operations taking into account the corresponding domain. Symbolic models, in K, and dynamical models, in S, are complementary in the sense that, neither type of model is reducible to the other. Both are necessary for understanding the whole system including classification schemas and dynamic processes.

The first-order wave equation, WE, is formally similar to the model (3.1) extensively applied in different domains by chemical engineers. For this reason the methodology based on the wave equation, WE, may be considered as a kind of artificial chemical engineering. It may be related to chemical engineering as the artificial chemistry is related to chemistry (Dittrich et al. 2001) or artificial life to natural life.

The physical, biological or engineering domains offer inspiration for the artificial domains, both for calculus and for artifacts.

3.3 Kinetic Model

For V=0 the first-order wave equation, WE, reduces to the kinetic model:

$$\frac{\partial Y}{\partial T} \oplus Q(Y) = 0 \tag{3.6}$$

The solution in GF (2) is presented here for illustration purposes. In GF (2), "0" denotes the null element. The real product and the sum were translated towards GF (2) operations.

Suppose that the rate of pattern recognition, Q, is the constant expansion denoted by

$Q = q_0 q_1 q_2 \ldots q_j$ or $Q = (q_j)$.

The solution similar to Euler solution for differential equations will be:

$$Y(T) = Y(0) \oplus Q \otimes T \tag{3.7}$$

Recall that the sum $\oplus$ is equivalent to the dyadic difference.

Suppose that, Y (0) = 1. In this case the solution of the first-order wave equation, WE, for different Q is Y (T, Q) and is shown in Table 3.3.

The detailed equations for m=0 are:

$$\frac{\partial y_0}{\partial t_0} \oplus q_0 = 0 \tag{3.8}$$

$$y_0(0) = f_0 \tag{3.9}$$

Table 3.3 Kinetic model, m=0

Q\T	0	1
0	1	1
1	1	0

Table 3.4 Kinetic model, modified, m=0

Q\T	0	1
0	1	1
1	1	-1

Denote, the resulting “0” by”-1”with the same logical signification, for instance “no”. Table 3.4 replaces Table 3.3.

Suppose that Y, T, Q are vectors with two components: $Y=y_0y_1$, $T=t_0t_1$, $Q=q_0q_1$ and $F=f_0f_1$.

This corresponds to two conditioning levels. The first-order wave equation, WE, reduces in fact to two similar equations, one for each level. For m=0 the model is given by equation (3.8) with initial condition (3.9).

For m=0 and m=1 a new equation and initial condition should be added:

$$\frac{\partial y_1}{\partial t_1} \oplus q_1 = 0 \tag{3.10}$$

$$y_1(0) = f_1 \tag{3.11}$$

The fact that one equation in K is replaced by two differential equations, one for each conditioning level, outlines one of the differences between models in K and in S.

Consider the initial condition:

$$Y(Z, 0) = F(Z) = f_0 \times f_1 \tag{3.12}$$

This means that the tensor product is interpreted as a categorical product, denoted by “$\times$”. The selection of tensor product is not limited to this (Iordache 2009).

The solution of the model will be:

$$Y(T) = y_0 \times y_1 \tag{3.13}$$

Specific case $f_0= f_1$, $q_0= q_1$, implies $y_0= y_1$.

Table 3.5 shows the product solution.

Table 3.5 Kinetic model, product $Y(T) = y_0 \times y_1$

1	1	1	1
1	-1	1	-1
1	1	-1	-1
1	-1	-1	1

This represents the Walsh-Hadamard, WH, matrices in DOE. With more coordinates in Y, T, Z it is possible to obtain Walsh matrices with 8, 16, 32, and so on elements.

Table 3.6 Kinetic model, product Y (T) = $y_0 \times y_1 \times y_2$

1	1	1	1	1	1	1	1
1	-1	1	-1	1	-1	1	-1
1	1	-1	-1	1	1	-1	-1
1	-1	-1	1	1	-1	-1	1
1	1	1	1	-1	-1	-1	-1
1	-1	1	-1	-1	1	-1	1
1	1	-1	-1	-1	-1	1	1
1	-1	-1	1	-1	1	1	-1

Table 3.6 shows the three conditioning levels solution for $Y=y_0y_1y_2$ and $T=t_0t_1t_2$ a Walsh–Hadamard DOE matrix.

It was considered that $f_0= f_1= f_2$, $q_0= q_1= q_2$, and this imposes $y_0= y_1= y_2$.

Walsh series as solution of differential equations in dyadic field have been obtained by (Iordache 2009).

3.4 Differential Posets

Differential posets are partially ordered sets of interest for the study of posets, lattices and polytopes. They have been introduced by Stanley in 1988 and since generalized in various different ways (Stanley 1988, Fomin 1994).

A partially ordered set P, or a poset for short, is a pair P=(S, $\leq$) of a set S together with an order relation $\leq$ which satisfies the following conditions for all x, y, z $\in$ S:

- $x\leq x$ (reflexivity)
- $x\leq y$ and $y \leq x$ imply x=y (anti-symmetry)
- $x\leq y$ and $y \leq z$ imply $x\leq z$ (transitivity)

A lattice is a poset in which any two elements have a unique supremum, the elements' least upper bound, called their join, and an infimum, the greatest lower bound, called their meet. A polytope is a set of vertices in R^n and their convex hull.

Lattice and polytopes play a key role in many modern fields of research, such as algebraic geometry, combinatorics, physics and engineering.

An element x is minimal if there is no y such that y<x and x is maximal if there is no y such that x<y.

Given a poset P and elements x, y$\in$ P, we say that y covers x if x < y and there is no z such that x < y< z. Then y is a cover of x and x is covered by y and we denote this relationship by $x \prec y$ or $y \succ x$.

For many posets, all order relations follow from the cover relations and transitivity.

A poset with this property is called locally finite.

Every locally finite poset has a naturally associated Hasse diagram. This is a graph whose vertices are elements of the posets and whose edges denotes cover relations, where if $x \prec y$ we draw x below y.

If x and y are elements of the same poset P and $x \leq y$ or $y \leq x$ we say that x and y are comparable.

A chain in a poset P is a set of elements of P which are pairwise comparable while an anti-chain is a set of points which are pairwise incomparable.

A locally finite poset is graded if $P = \cup_n P_n$ is the disjoint union of antichains P_n, indexed by a set of consecutive integers, such that $x \in P_n$ and $y \succ x$ imply $y \in P_{n+1}$ and all minimal elements belong to the same P_i.

Young lattices and Fibonacci-Young are two of the most studied differential posets (Lewis 2007).

Given a non-negative integer n, a partition of n is a finite nonincreasing list of positive integers $\lambda = (\lambda_1, \lambda_2, \ldots, \lambda_k)$ such that $\lambda_1 + \lambda_2 + \ldots, + \lambda_k = n$.

We denote this by $\lambda \vdash n$. An order on the partitions is defined as follows: given two partitions $\lambda = (\lambda_1, \lambda_2, \ldots, \lambda_k)$ and $\mu = (\mu_1, \mu_2, \ldots, \mu_j)$, $\lambda \leq \mu$ if and only if $k \leq j$ and $\lambda_i \leq \mu_i$ for $1 \leq i \leq k$. The poset composed of all integer partitions ordered in this way is known as Young's lattice and is denoted by Y.

Young's lattices may be represented as a graph whose vertices are partitions and two vertices λ and μ are adjacent if and only if $\lambda \leq \mu$ or $\mu \leq \lambda$ in the partial order of Y.

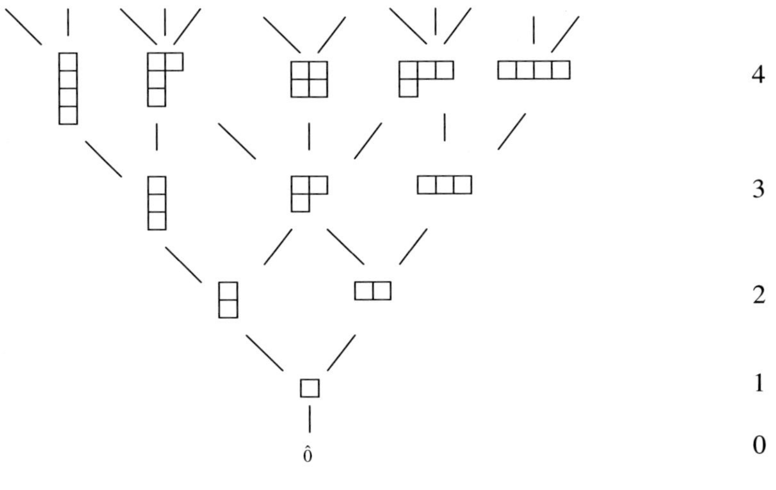

Fig. 3.1 Young lattice

Fig. 3.1 shows a Young's lattice.

Observe that there are five partitions of 4: (4), (3,1), (2,2), (2,1,1) and (1,1,1,1).

A poset P is differential if it satisfies the following three axioms:

DP1. P is locally finite and graded with a unique minimal element denoted by $\hat{0}$.

DP2. If $x \neq y$ are two elements of P and there are k elements of P covered by both x and y, there are exactly k elements of P which cover both x and y.

DP3. If $x \in P$ covers k elements of P then x is covered by exactly k+1 elements of P.

Given any poset P, we may define a vector space: $CP = \oplus_{x \in P} Cx$ of finite linear combinations of elements of P with complex coefficients where no additive relations hold among elements of P.

If in addition P is locally finite and each element of P is a member of only finitely many cover relations we may define two linear transformations U and D on CP as follows:

For $x \in P$,

$$Ux = \sum_{x \prec y} y \quad ; \quad Dx = \sum_{y \prec x} y \tag{3.14}$$

Both U and D should be extended to all of CP by linearity.

The study of the transformations U and D provides information concerning various paths in the Hasse diagrams, the so-called Hasse walks.

Basically Ux keeps track of all possible steps up in the Hasse diagram from x and Dx keeps track of all the steps down in the Hasse diagram.

The behavior of certain combinations of U and D steps on an arbitrary $x \in P$ is of interest for differential posets.

Observe that:

$$UDx = \sum_{y \prec z, y \prec x} z \quad ; \quad DUx = \sum_{z \prec y, x \prec y} z \tag{3.15}$$

It follows that DU-UD=I if and only if P is differential (Stanley 1988).

Moreover for a differential poset P we have:

$$DU^n = nU^{n-1} + U^n D \; ; \; D\hat{0} = 0 \tag{3.16}$$

Thus the action of D on U has a resemblance to that of a differential operator.

This explains the name of differential poset.

Several enumerative results on the class of differential posets have been derived using partial differential equations.

A generalization of differential posets is that of dual graded graphs (Fomin 1994).

A graded graph is a triple G= (P, ρ, E) where:

- P is a discrete set of vertices
- ρ: $P \rightarrow Z$ is a rank function
- E is a multiset of arcs (x, y) where $\rho(y) = \rho(x)+1$

The set $P_n=\{x: \rho(x) = n \in Z\}$ are called levels of G.

As for the differential posets the down and up operators can be defined.

Let G= (P, ρ, E) be a graded graph. Linear operators U and D are defined by:

$$Ux = \sum_{(x,y)\in Z} m(x,y)y \quad ; \quad Dy = \sum_{(x,y)\in Z} m(x,y)x \tag{3.17}$$

Here m (x,y) is the multiplicity of the edge (x,y) in E.

Let G_1= (P, ρ, E_1) and G_2= (P, ρ, E_2) be two graded graphs with a common set of vertices and a common rank function.

The oriented graded graph G= (G_1, G_2) = (P, ρ, E_1, E_2) is then the directed graded graph on P with edge in E_1, directed upwards and edges in E_2, directed downwards.

The down and up operators associated with the graph G= (G_1, G_2) are defined by:

$$Ux = \sum_{(x,y)\in E_1} m_1(x,y)y \quad ; \quad Dy = \sum_{(x,y)\in E_2} m_2(x,y)x \tag{3.18}$$

Here $m_i(x,y)$ denotes the multiplicity of (x, y) in E_i.

Let (G_1, G_2) be an oriented graded graph such that:

- It has a zero $\hat{0}$
- Each rank has a finite number of elements

Let r be a positive integer.

Then, G_1 and G_2 are said to be r-dual as operators in G= (G_1, G_2) if:

$$DU\text{-}UD=rI \tag{3.19}$$

If G_1 and G_2 are r-dual graphs, we call the pair (G_1, G_2) an r-dual graded graph.

A concept related to derivative poset is that of derivative complexes (Babson and Chan 2000).

Fig. 3.2 shows examples of derivative complexes.

If K is cubical poset, we can define a new cubical poset NK, with elements the ordered pairs (b, c) ∈ K x K such that the join of b and c covers both, while b and c have no meet.

Thus b and c are opposite facets of their join. The partial order on NK is the partial order on K taken component-wise.

Let ε: NK → NK denote the involution ε (b, c) = (c, b)

Then the derivative complex of K is the quotient poset DK=NK/ε.

Observe that NK and DK are both cubical posets and NK is a double cover of DK.

An element {b, c} = (b, c)/ε ∈ DK corresponds to a slice through the interior of the join of b and c, parallel to b and c. An element (b, c) ∈ NK corresponds to the side of {b, c} ∈ DK which faces b in K.

The NK and DK definitions are illustrated in Fig. 3.2.

Recall that a 1-D polytope is a line segment, a 2-D polytope is a polygon and a 3-D polytope is a polyhedron. Higher dimensional polytopes have been studied despite the fact that they are hard to illustrate (Ziegler 1995).

A cubical d-polytope is a d-dimensional convex polytope all of whose boundary faces are combinatorially equivalent to cubes.

The f vector of a cubical complex K is the vector f (K) = $(f_0, f_1,\ldots, f_{d-1})$ or $\sum_i f_i t^i$, where f_i denotes the number of i-dimensional faces in K.

Babson and Chan proved that:

$$\frac{d}{dt} f(K,t) = f(DK,t) \tag{3.20}$$

Moreover D and N act as derivations with respect to product and disjoint union, that is:

$$D (K_1 \times K_2) = (DK_1 \times K_2) \cup (K_1 \times DK_2) \tag{3.21}$$

Thus the action of D has a resemblance to that of a derivative and this explains the name of derivative complex.

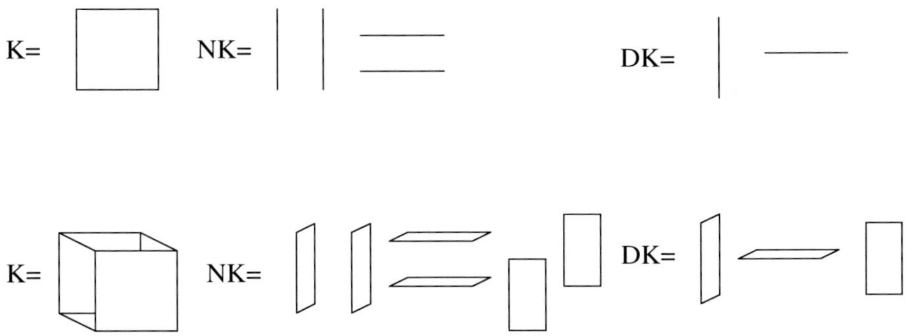

Fig. 3.2 Examples of derivative complexes

Examples of successful collaboration of the posets, lattices, polytopes and differential equations are due to Buchstaber (Buchstaber 2008). The approach is based on a differential ring of combinatorial polytopes. This allows applying the theory of differential equations to the study of polytopes. As an application, Buchstaber described explicitly the generating functions of important families of graph associahedra.

3.5 Doubling and Contracting

The operators U and D keeping track of the up or down steps for differential posets modify just one cell of a Hasse diagram.

A generalization of the differential poset concept is to consider the doubling and the contraction of intervals of lattices diagrams.

Constructive properties of lattice are of interest for the study of lattice boundedness.

A lattice is bounded if it can be constructed, starting with the one-element lattice, by applying a finite sequence of a simple operation called interval doubling (Day 1970).

This operation assigns to a poset P and an interval I, a new poset P′ = P[I] by doubling in P the interval I, that is by replacing I in P with the direct product by a two-element lattice.

Denote by "+" the disjoint set union.

The interval doubling construction is defined as follows (Caspard et al. 2004):

Let (P, ≤) be a poset and I ⊆ P an interval of P. Denote by B=({0,1} ≤) the two-element lattice where 0 <1. The poset P′ defined on the set (P-I) + (I x B) is denoted P′ = P[I] and is given by the following order: x′ ≤ y′ if and only if:

- x′, ∈ P-I and x′ ≤ y′ or
- x′∈ P-I , y′ =yi ∈ I x B and x′ ≤ y′ or
- x′=xi ∈ I x B, y′ ∈ P-I and x′ ≤ y′ or
- x′=xi ∈ I x B, y′ =yj ∈ I x B, x ≤ y and i ≤j in B.

A lattice L is bounded if there exist a sequence $B=L_1, \ldots, L_i, \ldots, L_p=L$ of lattices and a sequence $I_1, \ldots, I_i, \ldots, I_{p-1}$ such that Ii is an interval of L_i and $L_{i+1}=L_i[I_i]$ for every i<p.

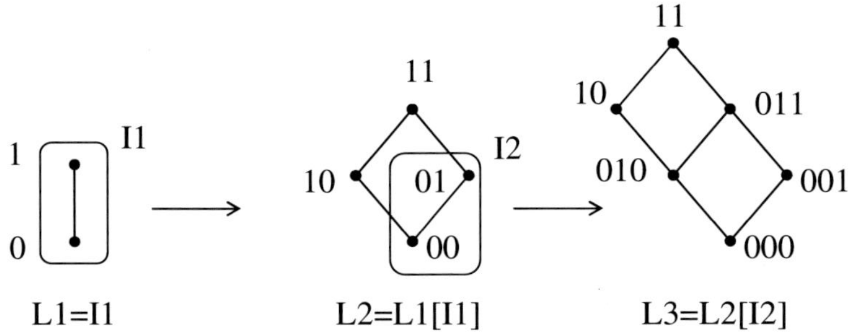

Fig. 3.3 Interval doubling

Fig. 3.3 shows a series of interval doublings starting with the two-element lattice B.

The lattices L1, L2 and L3 are bounded.

Since a bounded lattice is a lattice which can be constructed starting from B by a finite sequence of interval doublings, such a lattice is equally characterized by the fact that it can be contracted until B by an iteration of the operation opposite to the interval doubling. We can call this operation, interval contraction.

First we need to define the gluing conditions:

Let I be an interval of a lattice L, with I equal to the direct product of an interval I0 by B.

We denote by I1, the interval I- I0, isomorphic with I0. We say that I satisfy the gluing conditions if the two following conditions are verified:

- ∀ (y,x1,x0) ∈ (L-I1) x I1 x I0. (y<x1 → y≤x0)
- ∀ (z,x1,x0) ∈ (L-I0) x I1 x I0. (z>x0→z≥x1)

Let L be a lattice and I ⊆ L an interval of L. We say that I is contractible in L if the two following conditions hold:

• I is equal to the direct product of an interval I0 by B. I1 denotes the interval composed of the elements of I-I0.

• The gluing conditions are satisfied on I.

Fig. 3.4 illustrates the interval contracting.

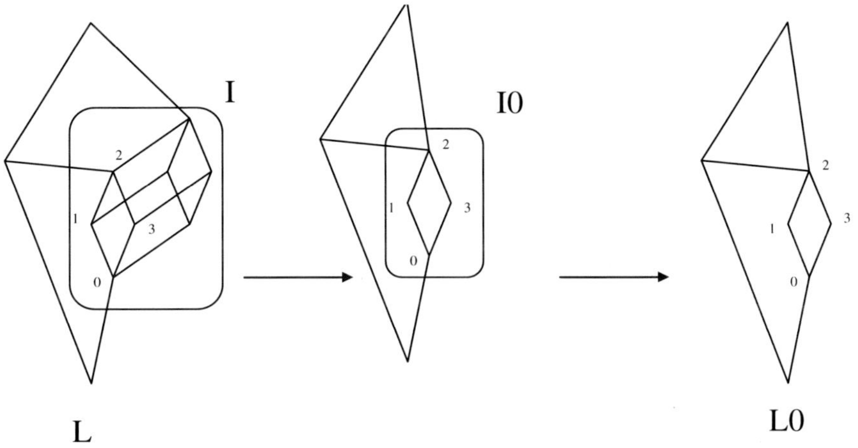

Fig. 3.4 Interval contracting

Let L be a lattice and I ⊆ L a contractible interval of L. We call contraction of I in L the operation of constructing a smaller lattice L0 by replacing I with I0 in L.

The contraction of an interval is the inverse operation to the interval doubling.

Fig. 3.4 shows the contraction of the interval I of the lattice L to the interval I0.

For the lattice L the interval I is contractible to I0.

Doubling and contracting allows obtaining a large variety of lattices generalizing the action of operators U and D defined for differential posets.

Doubling is associated to operator U while contracting corresponds to operator D.

3.6 Hopf Algebras

Important tools for the study of differential posets are the Hopf algebras (Sweedler 1969).

Hopf algebras were introduced, in the context of algebraic topology. The Hopf algebra is an algebra for which the dual space is also algebra so that the duality pairing is intertwined in a specific symmetric way. Examples range from group algebras, their duals, and universal enveloping algebras to deformations of such structures.

More recently the study of these algebras included applications in quantum groups, renormalization and non-commutative geometry.
Hopf algebra has been considered as a unifying framework for modeling several variants of multiplicative linear logic (Blute 1995).

Joni and Rota introduced Hopf algebra in combinatorics (Joni and Rota 1979).

Algebraic structures which are generated by a collection of constructors, like natural numbers generated by zero and a successor or finite lists and trees, are well known in engineering and computer science. Formally they are initial algebras. Induction is used for definition and for proofs for such structures. But there are also important dual coalgebraic structures which do not come equipped with constructor operations but with destructor operations also called observers, transition maps or mutators.

Spaces of infinite data including infinite lists and non-well-founded sets are of this kind. Dynamical systems, with a hidden state space to which the user has only a limited access via specified observer or mutator operations, are coalgebras of different types. Coinduction is the appropriate technique in this coalgebraic context, both as definition and proof principle. The latter involves the so-called bisimulations (Jacobs and Rutten 1997).

Coalgebras are structures that are dual, in the sense of reversing arrows to unital associative algebras. The axioms of unital associative algebras can be formulated in terms of commutative diagrams. Turning all arrows around, one obtains the axioms of coalgebras.

Hopf algebra is a structure that is simultaneously an unital associative algebra and a counital coassociative coalgebra, with these structures compatibility making it a bialgebra (Joni and Rota 1979).

Fig. 3.5 shows the Hopf algebra diagram.

To define the Hopf algebra, H as a k-vector space we need the following elements:

- Product μ: $H \otimes H \rightarrow H$
- Coproduct Δ: $H \rightarrow H \otimes H$
- Unit η: $k \rightarrow H$
- Counit ε: $H \otimes H \rightarrow H$
- Antipode map χ: $H \rightarrow H$

Here H denotes a vector space, k is a discrete field, id denotes the identity operator for H.

With these elements the diagram from Fig. 3.5 will commute.

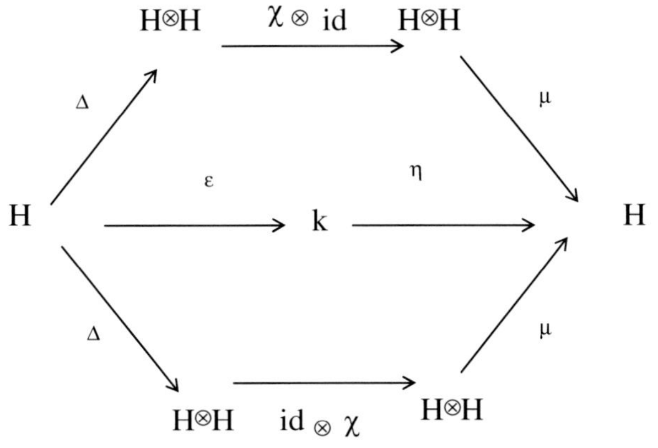

Fig. 3.5 Hopf algebra diagram

We refer to Hopf algebra H with antipode χ.

The coproduct Δ is interpreted as producing sums of pairs of new objects and this explains the potential for combinatorics.

The definition of Hopf algebra is self-dual, as reflected in the symmetry of the diagram from Fig. 3.5. So, if we can define a dual of H, which is always possible if H is finite-dimensional, then this dual is automatically Hopf algebra.

For the self-evolvable systems the concept of dual Hopf algebra is of interest.

Algebra and coalgebra, integration and differentiation are dual concepts.

Fig. 3.6 shows the polytope including algebra and dual algebra.

Fig. 3.6 suggests that after the integration or algebraic way S→K1→K2→K3 we need to look at the differentiation or dual algebraic way K3′→K2′→K1′→S′.

Making use of the developments of the direct way may offer in a kind of symmetry-breaking results. This opens the road for dual Hopf algebras interpretation (Hivert et al. 2005, Nzeutchap 2006). If the two ways offer the same results that is, in the case of self-duality the described system may be evolvable but not self-evolvable.

The swinging from algebra to dual algebra is critical since the boundaries where creative research grows and new information is created consist of parallel tendencies of integration and differentiation. The Self describes the interaction of the two algebras in duality relation.

Swinging method based on dual algebras has been applied in model evaluation and software correcting (Padawitz 2000, Jacob and Rutten 1997).

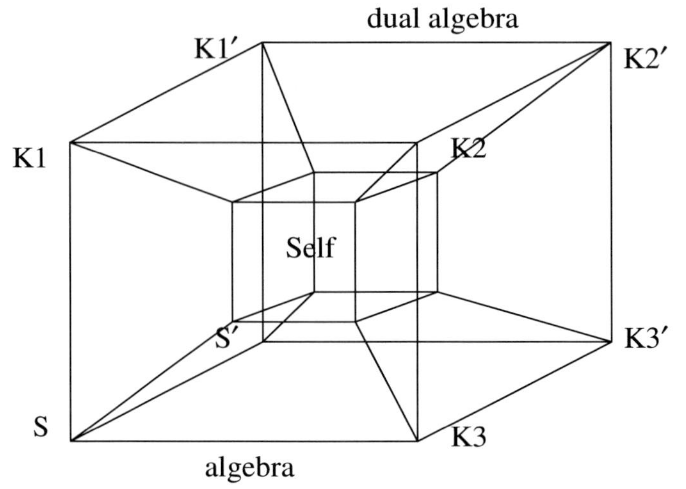

Fig. 3.6 Polytope for algebra and dual algebra

3.7 Differential Categories

Differentiation and differential models have been among the most fundamental tools for scientists and engineers.

Recently several attempts have been made to abstract these notions including approaches based on geometry, category theory and logical intuitions.

It was observed that the obtained new types of models are formally similar to the classical ones. They capture the intuitions from the ordinary calculus since we have calculus rules of differentiation expressed algebraically, formally similar to the classical ones. The new models represent an integral part of what it means to be differentiable.

Attempts to build categories of smooth structures have been suggested and are applied to the general notion of differentiation.

Synthetic differential geometry and linear logic are significant sources of inspiration for differential categories.

According to synthetic differential geometry SDG, an infinitesimal quantity can be taken to be a straight micro segment just long enough to have a slope but too short to bend (Kock 2006). It is an entity possessing location and direction without magnitude, intermediate in nature between a point and a straight line. In a smooth world any interval is indecomposable in the sense that it cannot be split in any way whatsoever into two disjoint nonempty parts.

The SDG provides the conceptual background for development of a mathematically based theory of potentiality and tendency. In conventional approaches, the life trajectory of actual items is characterized by the specific direction that it assumes at any one of its points and by the range of possibilities they have. On the other hand linelets and wavelets considered in SDG are too small to have either probabilities or directions. Instead, they have potentiality and tendency.

The SDG reasoning deals with space forms in terms of their structures that are the basic geometric and conceptual constructions that can be performed on them. The SDG constructions are morphisms which constitute the base category in terms of which we work, the space forms themselves being objects of it. This category is Cartesian closed.

A Cartesian category is a category with a terminal object and for any two objects in the category for instance A, B, the objects and mapping of the categorical product A x B are in the category. The object $\top$ of a category is called a terminal object if for every object in that category there is a unique map from it to $\top$. A category is Cartesian closed if it has all finite products and for any pair of objects say A and B there exists an exponential or map object A^B, the collection of all maps A→ B.

SDG reasoning is based on a category over a natural base topos (Baez 2006). Depending on the nature of the subject under consideration, the corresponding natural geometric form of the objects determines the natural base topos and its logic.

The objects of physics and chemistry have their own geometric form and corresponding logic. If the objects of the theory have a constant and crisp geometric form, we may use classical logic but if the geometric form is variable and fuzzy then we have to use a non-classical more flexible logic, for example the constructive logic.

Another source of inspiration for differential categories is the linear logic (Girard 1987). This is a resource sensitive logic or logic of actions and is based on categorical semantics.

The whole system of linear logic can be considered as an attempt to reconcile the systems of classical logic with the quest for constructive proofs that had led to constructive logic.

Recent developments in the model theory of linear logic have uncovered a variety of models with differential structure.

The development of formal systems for proving computational properties of programs constitutes an important research area of contemporary computer science. Among the aspects needing to be checked, one of the most important is the amount of resources a program will need during its execution. Resources to be bounded can be of different kinds, for instance memory space, or the non-replicable data.

A formal system modeling this idea of resource consumption was presented in Girard's quantitative semantics studies. This semantics establishes an analogy between linearity in the sense of computer science (programs using arguments exactly once) and algebraic linearity (commutation of sums and products with scalars), giving a new mathematically interpretation of resource consumption.

Ehrhard and Regnier designed a formal programming language, called the differential λ-calculus that has a unique kind of argument but two kinds of applications: the usual one and a linear algebra application (Girard 1987, Ehrhard and Regnier 2003). This linear substitution operation can be seen as a formal derivative. Such a syntactic derivative operator can be fruitfully used to increase control over programs executed in environments with bounded resources.

Differential proof nets contain a graph-theoretic syntax for linear logic extended with a differential operator on proofs (Ehrhard and Regnier 2006).

The differential categories and the differential Cartesian categories were defined on the basis of the works on differential λ-calculus (Blute et al. 2006, 2009).

For these categories a derivative operator on morphisms is axiomatized by equations. Typical examples are categories of vector spaces with normed or topological structure allowing defining a notion of smooth maps and of their differentiation.

The notion of a differential category provides a basic axiomatization for differential operators in monoidal categories, which not only generalizes the work of Girard or of Ehrhard and Regnier but also captures the standard elementary models of differential calculus and provides a theoretical substrate for studying a number of non-standard examples.

The basic categorical structure underlying the proof theory of Girard's linear logic (Girard 1987) is a symmetric monoidal closed category equipped with a comonad (Seely 1989).

Monads and comonads are valuable concepts in category theory. To any canonical construction from one type of structures to another, an adjunction between the associated categories, will correspond. Adjoint functors are pairs of functors which stand in a particular relationship with one another. A functor can be left or right adjoint to another functor that maps in the opposite direction. A pair of adjoint functors typically arises from a construction defined by a universal property, and it can be seen as a more abstract and powerful view on universal properties. If F and G represent a pair of adjoint functors, with F left adjoint to G right adjoint, then the composition G∘F will be a monad. The categorical dual of monads, F∘G, will be a comonad. Every adjunction gives rise to a monad. A monad is a functor from a category to itself, in other words an endofunctor. In general, the adjunctions relate categories of different natures. The monad theory tries to capture what is that adjunction preserves. The monads generalize closure operators on posets to arbitrary categories.

The structure necessary to support differentiation is an additive, monoidal category with a coalgebra modality. The morphisms in a differential category should be thought of as linear maps with maps in the category being the smooth maps (Blute et al. 2006).

Denote by ⊗ the multiplicative *and*, by 1 the unit, by ! the exponential modalities called: *of course* or *bang*.

A comonad (!, δ, ε) on an additive symmetric monoidal category, is a coalgebra modality in case each object !X comes equipped with a natural coalgebra structure given by:

$$\Delta: !X \rightarrow !X \otimes !X\ ;\ e: !X \rightarrow \top$$

Here ⊤ denotes the tensor unit. This data must satisfy some basic coherence relations (Blute et al. 2006).

The axioms for differential categories are:

D.1 Constant maps:
$D\,[e_A]=0$

Here e_A denotes the constant map.

D.2 Product rule:
$D\,[\Delta\,(f \otimes g)] = (1 \otimes \Delta)\ a_{\otimes}^{-1}(\,D[f] \otimes [g] + (1 \otimes \Delta)\, a_{\otimes}^{-1}(c_{\otimes} \otimes 1)\ a_{\otimes}\,(f \otimes D[g])$

Here f: !A→ B, g: !A→ C, and $a_{\otimes}$, $c_{\otimes}$ are the associativity and commutativity isomorphisms.

D.3 Linear maps:
$D\,[e_A\, f] = (1 \otimes e_A)\ u_{\otimes}\, f$

Here f: A→B and $u_{\otimes}$ is the unit isomorphism.

D.4 The chain rule:
$D\,[\delta\ !f\ g] = (1 \otimes \Delta)\ a_{\otimes}^{-1}(D[f] \otimes \delta\ !\ f)\ D[g]$

Here f: !A→ B!, and g: !A→ C!

Observe that according to (D.1) the derivative of a constant is null, and the derivative of a map which is linear is constant according to (D.3). The derivative of the composite of two functions is the derivative of the first function composed with the derivative of the second function at the value produced by the first function as shown by (D.4).

A significant step was to develop an axiomatization which directly characterizes the smooth maps, in other words, to characterize the structure of differential categories directly (Blute et al. 2009). This leads to the notion of a Cartesian differential category. This notion embodies the multi-variable and multi-scale differential calculus.

The basic structure needed for Cartesian differential categories is simpler than is needed for differential categories. It is a left additive category with finite products.

A Cartesian left additive category is a left additive category with products such that the structure maps π_0 π_1 and Δ are additive and that whenever f and g are additive the product is additive.

The axioms for Cartesian differential categories are:

CD.1 $D[f+g] = D[f] + D[g]$; $D[0]=0$
Operators preserve additive structure that is the operator D is linear

CD.2 $D[f]\circ\langle h+k,v\rangle = D[f]\circ\langle h,v\rangle + D[f]\circ\langle k,v\rangle$ and $D[f]\circ\langle 0,v\rangle = 0$
This shows additivity in first argument

CD.3 $D[1]=\pi_0$, $D[\pi_0]=\pi_0\circ\pi_0$ and $D[\pi_1]=\pi_0\circ\pi_1$
Coherence maps are linear differential constant

CD.4 $D[\langle f,g\rangle]=\langle D[f], D[g]\rangle$
Operators preserve pairing, that is D behaves coherently with the product structure

CD.5 $D[f\circ g]= D[f] \circ \langle D[g], g\circ\pi_1 \rangle$
This corresponds to the chain rule

CD.6 $D[D[f]] \circ \langle\langle g, 0\rangle, \langle h, k\rangle\rangle =D[f]\circ \langle g,k\rangle$
Differentials D are linear in first argument.

CD.7 $D\ [D\ [f]\circ \langle\langle 0, h\rangle, \langle g, k\rangle\rangle\ = D\ [D\ [f]]\circ \langle\langle 0,g\rangle,\langle h,k\rangle\rangle$

This means that partial differentials commute.

In Cartesian differential category we obtain partial derivatives from the full ones by zeroing out the components on which the differentiation is not required.

Developments of the notion of differential categories have been proposed (Fiore 2007, Bucciarelli et al. 2010).

Observe that there is a general formalism of differential categories, but there are certain specific ways that, scientists and engineers may use these which turns out to have close analogues in different domains.

References

Babson, E.K., Chan, C.: Counting faces for cubical spheres modulo two. Discrete Math. 212, 169–183 (2000)

Baez, J.: Topos theory in a nutshell (2006), `http://math.ucr.edu/home/baez/topos.html`

Blute, R.F.: Hopf algebras and linear logic. Mathematical Structures in Computer Science 6(2), 189–217 (1996)

Blute, R., Cockett, J.R.B., Seely, R.A.G.: Differential Categories. Mathematical Structures in Computer Science 16, 1049–1083 (2006)

Blute, R.F., Cockett, J.R.B., Seely, R.A.G.: Cartesian Differential Categories. Theories and Applications of Categories 22, 622–672 (2009)

Bucciarelli, A., Ehrhard, T., Manzonetto, G.: Categorical models for simply typed resource calculi. Electronic Notes in Theoretical Computer Science 265, 213–230 (2010)

Buchstaber, V.M.: Ring of Simple Polytopes and Differential Equations. Proceedings of the Steklov Institute of Mathematics 263, 1–25 (2008)

Caspard, N., Le Conte de Poly-Barbut, C., Morvan, M.: Cayley lattices of finite Coxeter groups are bounded. Adv. in Appl. Math. 33(1), 71–94 (2004)

Day, A.: A simple solution to the word problem for lattices. Canad. Math. Bull. (13), 253–254 (1970)

Dittrich, P., Ziegler, J., Banzhaf, W.: Artificial chemistries-a review. Artificial Life 7(3), 225–275 (2001)

Ehrhard, T., Regnier, L.: The differential lambda-calculus. Theoretical Computer Science 309(1-3), 1–41 (2003)

Ehrhard, T., Regnier, L.: Differential interaction nets. Theoretical Computer Science 364(2), 166–195 (2006)

Fiore, M.P.: Differential Structure in Models of Multiplicative Biadditive Intuitionistic Linear Logic. In: Della Rocca, S.R. (ed.) TLCA 2007. LNCS, vol. 4583, pp. 163–177. Springer, Heidelberg (2007)

Fomin, S.: Duality of graded graphs. J. Algebraic Combin. 3, 357–404 (1994)

Girard, J.-Y.: Linear logic. Theoretical Computer Science 50, 1–102 (1987)

Harmuth, H.F.: Sequency Theory, Foundations and Applications. Academic Press, New York (1997)

Hivert, F., Novelli, J.-C., Thibon, J.-Y.: The algebra of binary search trees. Theoretical Computer Science 339, 129–165 (2005)

Iordache, O.: Evolvable Designs of Experiments Applications for Circuits. J Wiley VCH, Weinheim (2009)

Iordache, O.: Polystochastic Models for Complexity. Springer, Heidelberg (2010)

Iordache, O.: Modeling Multi-Level Systems. Springer, Heidelberg (2011)

Iordache, O., Corriou, J.P., Garrido-Sanchez, L., Fonteix, C., Tondeur, D.: Neural network frames applied for biochemical kinetic diagnosis. Comp. Chem. Engng. 17, 1101–1113 (1993a)

Iordache, O., Valentin, G., Corriou, J.P., Pons, M.N., Pethö, A.: Intermittent Interfacial Transfer. A Dyadic Model. Acta Chemica Hungarica, Models in Chemistry 1(130), 1–18 (1993b)

Iordache, O., Corriou, J.P., Tondeur, D.: Separation Sequencing Use of Information Distance. Canad. Journ. of Chem. Engng. 71, 955–966 (1993c)

Jacobs, B., Rutten, J.: A tutorial on (co)algebras and (co)induction. EATCS Bulletin 62, 222–259 (1997)

Joni, S.A., Rota, G.C.: Coalgebras and Bialgebras in Combinatorics. Stud. Appl. Math. 61, 93–139 (1979)

Kock, A.: Synthetic Differential Geometry, 2nd edn. Cambridge University Press, New York (2006)

Lewis, J.B.: On Differential Posets. Undergraduate Thesis. Harvard College (2007)

Nzeutchap, J.: Dual graded graphs and Fomin's r-correspondences associated to the Hopf algebras of planar binary trees, quasi-symmetric functions and noncommutative symmetric functions. In: International Conference on Formal Power Series and Algebraic Combinatorics, FPSAC, San Diego (2006)

Padawitz, P.: Swinging Types = Functions + Relations + Transition Systems. Theoretical Computer Science 243, 93–165 (2000)

Poli, R.: Three Obstructions: Forms of Causation, Chronotopoids, and Levels of Reality. Axiomathes 17, 1–18 (2007)

Rhee, H.K., Aris, R., Amundson, N.R.: First Order Partial Differential Equations II. Theory and applications of hyperbolic systems of quasilinear equations. Prentice-Hall, Englewood Cliffs (1989)

Seely, R.: Linear logic, *-autonomous categories and cofree coalgebras. Applications of categories in logic and computer science. In: Gray, J., Scedrov, A. (eds.) Contemporary Mathematics, vol. 92, pp. 371–382 (1989)

Sweedler, M.: Hopf Algebras. Benjamin Press (1969)

Stanley, R.: Differential posets. J. Amer. Math. Soc. 1, 919–961 (1988)

Ziegler, G.M.: Lectures on Polytopes. Graduate Texts in Mathematics, vol. 152. Springer, New York (1995)

Chapter 4
Informational Criteria

Abstract. Entropy and entropy production for multi-level systems are studied here with reference to informational aspects.

For the general PSM framework, new entropic criteria are proposed. The criteria correspond to different types of causation. Evolvability maximization and self-evolvability maximization role for self-integrative closure systems is emphasized.

4.1 Informational Entropy

Classification, categorification, separation are all based on similarity evaluation.

The initial step in quantifying the concept of similarity for objects, as for instance the chemical species in a mixture, is to list the most important structural elements or properties of the species.

For chemical mixture examples, binary bit string representations of molecular structure and properties, the so-called fingerprints are standard tools to analyze similarity (Willett 1998).

To every species in a mixture can be associated a vector the components of which take only two values "1" or "0" where "1" means the presence of a given structural element or property whereas "0" means its absence. For instance, "1" may corresponds to a high value of the hydrophilic character, whereas "0" corresponds to a low value or hydrophobic character.

Vectors associated to different objects are denoted by: $i = [i_1, i_2, \ldots, i_k, \ldots]$ where i_k are either "1" or "0".

A hierarchy of the structural elements or properties is required. For instance, it is considered that the property indexed by i_1 is more significant than the property indexed by i_2, this more significant than i_3, and so on in the order of coordinates in the associated vectors (Iordache et al. 1993b, Iordache 2010).

To any set of compounds or objects a similarity matrix is associated and to this an informational entropy. On this basis the components of the mixture may be selected.

A similarity index r_{ij}, between two different species $i= [i_1, i_2,..., i_k,...]$ and $j= [j_1, j_2,..., j_k,...]$, is defined as:

$$r_{ij} = \Sigma_k t_k \ (a_k)^k \ ; \ k=1,2,.... \tag{4.1}$$

Here: $0 \leq a_k \leq 1$ and $t_k=1$ if $i_k = j_k$, $t_k=0$ if $i_k \neq j_k$ for all k. The entire system is characterized by the matrix $R=[r_{ij}]$. The similarity index should possess the natural properties of reflexivity ($r_{ii} = 1$) and of symmetry ($r_{ij}=r_{ji}$). This definition assigns a coefficient of weight a_k to any property involved in the description of the species i and j provided the Boolean values i_k and j_k are the same for these two objects.

The fact that the relation described by r_{ij} is reflexive and symmetric allows a partition of the set of components in classes that are not necessarily disjoint. A class consists of a number of similar species gathered together. To limit the study to partition into disjoint classes the defined similarity must be transitive that is: $\min_k (r_{ik}, r_{kj}) \leq r_{ij}$. The procedure to ensure transitivity is that the classification algorithm starts from the stable matrix of similarity. To obtain such a stable matrix, the sequence R, R(2),..., R(k),...with R(2)=RoR and R(k)=R(k-1)oR is calculated. The composition rule "o" is given by:

$$(RoW)_{ij} = \max_k [\ \min (r_{ik}, w_{kj})\ ] \tag{4.2}$$

Here $R= [r_{ij}]$, $W= [w_{ij}]$ are two arbitrary matrices of the same type. The composition equation calculates the (i,j)th element of the matrix RoW. It consists in taking the smallest of the two elements r_{ik} and w_{kj}, for a given row i of R and a column j of W, then repeating the procedure for all k and selecting the largest of all such resulting elements. There exists an integer n, such that from n on, the matrix is stable to the composition rule "o" so that R(n)=R(n+1) and so on. We refer to n as the number of stabilization steps.

The elements of the stable similarity matrix R(n) verify symmetry, reflexivity and transitivity.

Denote by r_{ij} (n) the elements of the stable matrix R(n). The partition in classes is established on the base of the degree of classification T with $0 \leq T \leq 1$. The classification rule is the following: two species i and j are assigned to the same class if r_{ij} (n)$\geq$ T. Applying the rule, the set of classes at the degree of classification T is obtained. For T=0, a unique class results including all species, whereas for T=1 each class includes only one species. When T varies from 0 to 1, different sets of classes arise. Actually a new set of classes arises every time T crosses the value of one similarity index r_{ij} of the matrix R. In this way a general tree of classes is built, which is the expected schema. The class of i, denoted î, is the set of species j which satisfies the rule: $r_{ij} \geq T$.

The similarity matrix of classes $\hat{R}$ is constructed as follows:

$$\hat{R}_{\hat{i}\hat{j}} = \max(r_{wu}); w \in \hat{i}, u \in \hat{j} \tag{4.3}$$

Here w designates any index of species belonging to the class of $\hat{i}$ and similarly u any index referring to the class of $\hat{j}$.

To any similarity matrix R, the informational entropy H(R) is associated:

$$H(R) = - \Sigma\, r_{ij} \ln r_{ij} - \Sigma\, (1- r_{ij}) \ln (1- r_{ij}) \tag{4.4}$$

This expresses the quantity of information associated to the matrix R.

The defined entropy is a measure of imprecision in classifying.

To compare two similarity matrices R= $[r_{ij}]$ and W= $[w_{ij}]$ a distance DD was introduced:

$$DD(R, W) = -\Sigma\, r_{ij} \ln (r_{ij} / w_{ij}) - \Sigma(1- r_{ij}) \ln ((1- r_{ij})/ (1-w_{ij})) \tag{4.5}$$

The distance measure the discrepancy between two similarity matrices and associated classifications.

There are different definitions of similarity based on the definition of distance between two vectors.

The Minkowski distance is:

$$d_{ij} = (\Sigma_k (i_k-j_k)^k)^{1/k} \tag{4.6}$$

If k=2 this gives the Euclidean distances. The Manhattan or city-block distance is also of interest. This is defined as:

$$d_{ij} = \Sigma_k \left| i_k-j_k \right| \quad 4.7)$$

It is equivalent to the number of settings, which are different in the vectors i and j.

The Hamming distance is defined by:

$$d_{ij} = \Sigma_k \, XOR\, (i_k, j_k) \tag{4.8}$$

XOR is the exclusive or logical function. For Boolean vectors, the Manhattan and the Hamming distances are equivalent.

The Tanimoto distance is defined by:

$$d_{ij} = \Sigma_k \, AND\, (i_k, j_k)/OR\, (i_k, j_k) \tag{4.9}$$

AND and OR are logical functions.

Another distance is:

$$d_{ij} = \Sigma_k \, t_k\, (a_k)^k \tag{4.10}$$

With $0 < a_k<1$, a constant, $t_k =1$ if $i_k = j_k$, and $t_k=0$ starting from the first k with: $i_k \neq j_k$.

If the comparison between i and j stops at the first difference in coordinates is detected and $a_k = 0.5$ the distance is ultrametric.

Distances are measures of dissimilarity. Once the distance is established, similarity indices may be obtained using:

$$r_{ij} =1/(1+d_{ij}) \tag{4.11}$$

$$r_{ij} =1-d_{ij}/\max (d_{ij}) \tag{4.12}$$

Observe that $r_{ij} <1$ and that $r_{ii} =1$.

4.2 Entropy Criteria for Self-Evolvability

Informational entropy criteria proved to be useful for evolvability studies (Iordache 2011). The starting point was that the integrative closure as shown by evolvable systems involves different kinds of causality.

The considered hierarchy of causes was: material→formal→efficient→final.

A hierarchy that places efficient cause before formal cause that is: material→ efficient→ formal→final has been considered by several authors (Kineman 2010).

Different types of causes may be associated to different levels of the general framework associated to PSM and to different entropy criteria.

At successive levels, successive derivatives or differences of entropy became very low, close to null. This suggests correlating the criteria of maximum entropy ME, to mechanical causation, the criteria of maximum production of entropy MEP, to the formal causation, the criterion of maximum production of entropy production MPEP, to efficient causation and the evolvability maximization EM, to final causation (Iordache 2011).

Fig. 4.1 illustrates the integrative closure and the entropy criteria. It refers to what is supposed to happen in evolvable systems.

Material causation, extensively used in chemistry and biology, seems more linked, to what Peirce describes as firstness.

Formal causation studies are well developed in second-order cybernetics.

Efficient causation manifests when the regularities become significant for interactions through stable patterns. Efficient causation concept is allied, to what Peirce describes as secondness.

As the efficient causation we are looking for the criteria of maximum production of entropy production (MPEP) correlated to a third derivative of the entropy.

We may also correlate MPEP with a tentative fourth law of thermodynamics, in which the workspace of the biosphere expands, on average, as fast as it can in the co-constructing biosphere (Kauffman 2000). By as fast as it can, Kauffman means something like the edge of chaos. Faster than that, it cannot sustain itself. Slower is not advantageous since if there is the possibility of going faster, the faster ones become selected.

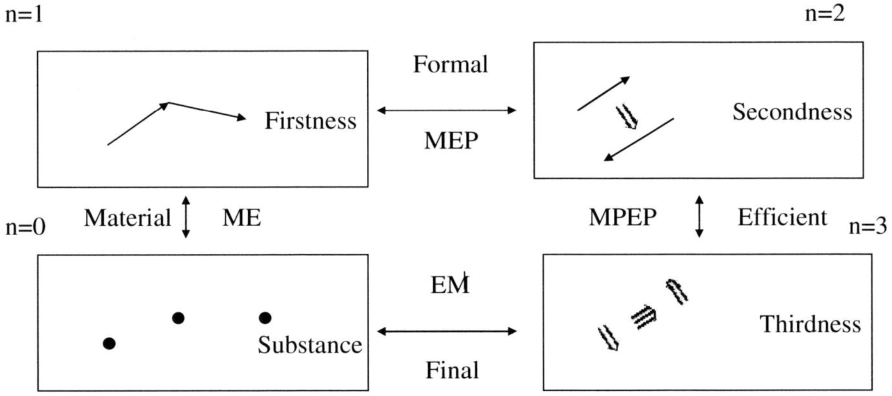

Fig. 4.1 Integrative closure and entropy criteria

MPEP criterion is linked to the maximum power principle due to Odum (Odum 1988, 1995). Natural systems tend to self-organize in multiple scales by using energy and materials to build structures, which function to feedback and amplify their capture and use. Since Lotka classical contribution, this autocatalytic relationship has been considered to be a basic organizing principle in the emergence of life and the overall organization of nature. Odum considered that a system's capacity to prevail in evolution was related to its capacity to capture useful power. The maximum power principle can be stated as follows: during self-organization the system designs develop and prevail that maximize power intake, energy transformation, and those uses that reinforce production and efficiency. Self-organizing systems disperse energy faster, maximizing the rate of entropy production by developing autocatalytic dissipative structures.

MPEP is allied to a promising candidate for an information concept that has been developed in ecology as the so-called average mutual information, a measure for how well organized or determinate a configuration of ecological relationships appears. Ulanowicz has developed this concept further to a concept of ascendency that represents the coherent power a system could bring to bear in ordering itself and the world around it (Ulanowicz 1997).

A proposal was to use informational distance DD, as a measure of mutual information, to run simple MPEP calculus (Iordache 2011).

Fig. 4.1 shows that the basic level n=0 refers to states distribution and for these, ME criteria ensure entropy increasing.

The level n=1 refers to processes and for these MEP criteria it is a statement about possible trajectories and looks to the most typical trajectory. The 1-arrows associated to the level n=1 are trajectories or paths corresponding to 1-categories.

The level n=2 refers to processes of processes and to 2-categories. MPEP would govern the interaction between trajectories. The 2-arrows are ways of sweeping from one trajectory to the other.

The 3-arrows are defined between pairs of 2-arrows and consist of ways of interpolating between the sweepings from one trajectory to the other. This level refers to 3-categories.

The final causation is the critical step for integrative closure. For this step, the criteria of evolvability maximization, EM, were proposed.

The EM involves the embodiment into substance S and may be correlated in part, to the concept of maximization of the mutual information as described by Sporns and Lungarella (Sporns and Lungarella 2006). They demonstrated how the maximization of the information structure of the sensory states experienced by embodied and situated agents might lead to the development of useful behavioral skills.

Such studies suggested the formulation evolvability maximization, EM, criterion for the step correlating the level n=3, thirdness to the level n=0, in integrative closure.

The final causation concerns the goal to be achieved by the system. This is the level where the goals are influencing the results. Embodiment and empowerment are necessary marks. Evolvability is a measure of the informational efficiency for the integrative closure.

A generic approach to derive fundamental candidates for systemic drives from properties of the integrative closure loop emphasizes the system as an entity that is able to select its actions (Klyubin et al. 2005, Capdepuy 2007, Polani 2009). It considers the informational channel capacity between the system's action at a given time and the sensory inputs at later time. Intuitively this is a measure to which extent the system's actions could potentially influence its system, in such a way that the intensity of this influence can later be detected again by the system. The, empowerment, measures the system's power to change the environment and to be aware that it did so. This can be formalized by measuring the maximal mutual information that can possibly be introduced into the environment by a suitable distribution of actions (Klyubin et al. 2005).

Evolvability maximization, EM, criterion could be interpreted as follows: any evolvable system should poise itself in such a way as to be able to react in a most effective way to possible perturbations of its preferential state. The higher evolvability, the better is the possibility of the system to self-control the perturbations. This corresponds to a kind of cognitive behavior, to self-organization in integrative closure and creates systems with an interest to preserve their own organization. The open problem is that one needs to identify beforehand the variables whose stabilization is necessary for the particular system.

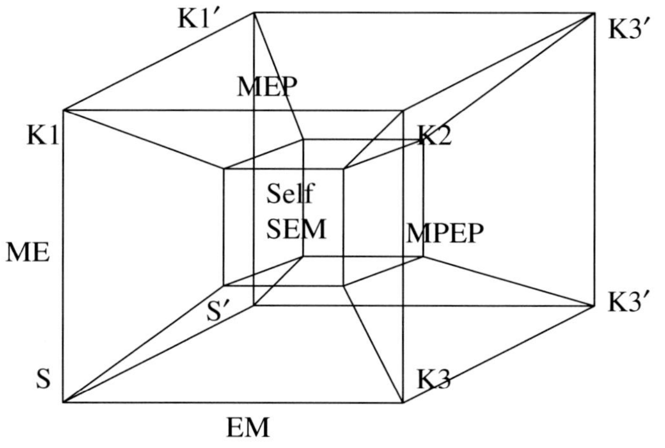

Fig. 4.2 Polytope for self-integrative closure and entropy criteria

Infotaxis concept may also be linked to EM criterion (Vergassola et al. 2007). Infotaxis is designed to work when the searcher must devise a strategy of movement based on sporadic cues, weak signals and incomplete information. Any search process can be thought of as acquisition of information on source location. For infotaxis, signs or partial information play a role similar to concentration in chemotaxis. Infotaxis is motion based on expected information gain. The infotaxis strategy locally maximizes the expected rate of information gain.

Fig. 4.2 illustrates the polytope for self-integrative closure and all the entropy criteria.

It refers also to what is supposed to happen in self-evolvable systems, that is, the self-evolvability maximization, SEM.

The self-evolvability takes into account a fifth causation supplementing the Aristotle's four causes. This supplementary cause, discussed by several authors, was identified in different ways as: pattern, anticipation, exemplary, idea, chance, or essence. Louie referred to the fifth cause as exemplary cause and described it as the potentiality that anticipates another actuality, the swinging between being and becoming of the formal cause of something else (Louie 2009).

The complementary roles of the being and becoming for self-evolvability should be emphasized. The swinging between being and becoming is the tool for innovative or creative systems that can autonomously find solutions to highly complex problems.

We may consider the fifth cause as the self-cause. It appears as a kind of meta-cause, putting together the other four types of causation.

SEM criterion would include an anticipation mechanism.

The anticipation follows the exemplars that are meaningful over time, to sustain the functions of organisms. The ability of a system to change into a different kind of system by changing its function in nature, its code, and thus to change its suitability for existence, is a new dimension of behavior that is more than a mechanism and more than an adaptive system.

The agility is another concept that may be useful for SEM criteria studies. Agility is defined as the ability of surviving and prospering in a competitive environment of continuous and unpredictable change by reacting quickly and effectively to changing environment. Agility supposes a rapid, proactive or anticipative adaptation of systems' elements to unexpected and unpredicted changes.

It is known that autocatalytic feedback is a significant route by which systems increase and maintain their self-evolvability. For modeling purposes this implies to make use of hypersets and non-well-founded probabilities (Iordache 2011).

An open problem for both EM and SEM criteria study is that one needs to identify beforehand the variables sustaining functioning and stabilization for the particular system.

A suitable SEM criterion will go beyond standard definitions of informational entropy. It cannot be based on conventional probability or similarity mechanisms only.

As suggested by Fig. 4.2 the swinging between upward causation way S→K1→K2→K3 and downward causation way K3′→K2′→K1′→S′ should be taken into account for self-evolvability. Considered as categorical duality this swinging corresponds to reversing the arrows directions.

Self-evolvability is based on both upward and downward causation models (Campbell and Bickhard 2011, Pattee 2000).

Making sense of upward and downward causation does not require the acceptance of a distinction between the higher and basic levels of organization. It is enough to assume a duality of modes of access, or modes of intervention.

If one intervenes at a higher level of organization, some effects of this action can then be detected by a mode of access specifically aimed at a lower level. This is downward causation. Conversely, if one intervenes at a lower level, some effects of this action can then be detected by a mode of access specifically aimed at a higher level of organization. This is upward causation. Sporns and Tononi (2007) made use of informational criteria to evaluate the coexistence of upward causation that is integration way with backward causation that is differentiation way for functional brain dynamics study.

Their attempt may be a source of inspiration for SEM informational criteria.

We are looking for SEM criterion only for systems outlining both ways the direct and the indirect way in their evolution. If there is no difference between these two ways we limit the search for criteria to EM, characterizing evolvability.

Table 4.1 summarizes the categorification aspects for entropy criteria

Table 4.1 Categorification for entropy criteria

Level	K0 (S)	K1	K2	K3	Self
-	n=0	n=1	n=2	n=3	n=4
Categories	0-category	1-category	2-category	3-category	4-category
Entropy Criteria	ME	MEP	MPEP	EM	SEM

Criteria as ME characterizes transition from n=0 to n=1, MEP that from n=1 to n=2, and MPEP the transition from n=2 to n=3.

EM criterion refers to the connection between level K3, that is n=3 and S, that is n=0. This is associated to integrative closure and evolvability. SEM refers to the connection between K3 and K3′, self-integrative closure and the emergence of self-evolvability.

References

Campbell, R.J., Bickhard, M.H.: Physicalism, Emergence and Downward Causation. Axiomathes 21, 33–56 (2011)

Capdepuy, P., Polani, D., Nehaniv, C.: Maximization of potential information flow as a universal utility for collective behavior. In: Proceeding of the 2007 IEEE Symposium on Artificial Life (CI-ALife 2007), pp. 207–213. IEEE Press, Piscataway, NJ (2007)

Iordache, O.: Polystochastic Models for Complexity. Springer, Heidelberg (2010)

Iordache, O.: Modeling Multi-Level Systems. Springer, Heidelberg (2011)

Iordache, O., Corriou, J.P., Tondeur, D.: Separation Sequencing. Use of Information Distance. Canad. Journ. of Chem. Engng. 71(12), 955–966 (1993b)

Kauffman, S.: Investigations. Oxford University Press (2000)

Kineman, J.J.: Relational Science: A Synthesis. In: Allen, J., Wilby, J. (eds.) Proceedings of the 54th Meeting of the International Society for the Systems Sciences, ISSS, Waterloo (2010)

Klyubin, A.S., Polani, D., Nehaniv, C.L.: Empowerment: A universal agent-centric measure of control. In: Proc IEEE Congress of Evolutionary Computation (CEC 2005), Edinburgh, UK, pp. 128–135 (2005)

Louie, A.: More than life itself: A synthetic continuation in relational biology. Ontos Verlag, Frankfurt (2009)

Odum, H.T.: Self-Organization, Transformity and Information. Science 242, 1132–1139 (1988)

Odum, H.T.: Self-Organization and Maximum Empower. In: Hall, C.A.S. (ed.) Maximum Power: The Ideas and Applications of H.T.Odum. Colorado University Press, Colorado (1995)

Pattee, H.H.: Causation, control and the evolution of complexity. In: Anderson, P.B., Christiansen, P.V., Emmeche, C., Finnemann, N.O. (eds.) Downward Causation, pp. 63–67. Aarhus University Press (2000)

Polani, D.: Information: currency of life? HFSP Journal 3(5), 307–316 (2009)

Sporns, O., Lungarella, M.: Evolving coordinated behavior by maximizing information structure. In: Rocha, L.M., Yaeger, L.S., Bedau, M.A., Floreano, D., Goldstone, R.L., Vespignani, A. (eds.) Artificial Life X, Proccedings of the Tenth International Conference on the Simulation and Synthesis of Living Systems, pp. 323–329. MIT Press, Cambridge (2006)

Sporns, O., Tononi, G.: Structural determinants of functional brain dynamics. In: Jirsa, V., McIntosh, A.R. (eds.) Handbook of Brain Connectivity, pp. 117–148. Springer, Berlin (2007)

Ulanowicz, R.E.: Ecology, the ascendant perspective. Columbia University Press, NY (1997)

Vergassola, M., Villermaux, E., Shraiman, B.I.: Infotaxis as a strategy for searching without gradients. Nature 445(7126), 406–409 (2007)

Willett, P.: Chemical similarity searching. J. Chem. Inf. Comput. Sci. 38, 983–996 (1989)

Chapter 5

Self-Evolvability for Physical and Chemical Systems

Abstract. Self-evolvability potentialities for physical systems are presented using differential models, operads and entropy criteria.

Schemas enumeration, separation trees, process synthesis, cyclic operations, dendritic growth, biochemical substrates for technical information processing, circuits and antennas are the studied systems.

5.1 Separation Schemas

The unit operations are the basic steps for chemical engineering processes. A process has several unit operations structured to obtain the desired product.

Separation operations transform a mixture of substances into distinct products. The separated products should differ in some properties, such as size, density, electric charge, solubility, volatility, affinity, reactivity and so on.

Centrifugation and sedimentation are based on density differences.

Chromatography separates dissolved substances by different interaction with a material.

Electrophoresis separates molecules based on their different interaction with a gel under an electric potential.

Distillation is convenient for mixtures of liquids with different boiling points. Drying removes liquid by vaporization.

The mechanical separations are favored due to the lower cost of the operations as compared to chemical separations. For systems that cannot be separated by purely mechanical means, chemical separation is the remaining solution. The operated mixture could exist as a combination of any two or more states as for instance, solid-solid, solid-liquid, solid-gas, liquid-liquid, liquid-gas, gas-gas, and solid-liquid-gas mixture.

Depending on the raw mix, various processes can be employed to separate the mixtures.

The processes of separation are described by separation schemas.

There exists a deep relationship between coherence studies in category theory and the identification of separation schemas for mixtures (Iordache 2010).

The term coherence covers in category theory what from a logical point of view would be called problems of completeness, axiomatizability and decidability.

Engineering domains may suffer from the lack of coherence for instance if properties and the corresponding databases are incompatible and different systems do not work together.

It should be noted that a concept of coherence in chemical engineering came out in the study of fixed-bed ion exchange and adsorption operations (Helfferich and Klein 1970). Major extensions were to multiphase systems in enhanced oil recovery, to chemical reactors and to process synthesis. The coherence concept in chemical engineering was linked to mass transfer models based on hyperbolic differential equations.

A general concept of categorical coherence providing an understanding of the difference between coherent and non-coherent situations may be based on the wave equation, WE and on differential categories.

Consider for example that the axiom to be imposed to possible separation sequence is the associativity. This means that, within a sequence of elements containing two or more of the same sequencing operations in a row, the order that the operations are performed does not matter as long as the sequence to be operated is not changed. Rearranging the parentheses in such a sequence will not change the sequencing general task.

The resulting associahedrons are studied in category theory as coherence conditions.

Suppose that there are four components and that the order of sequencing is imposed from start, by a heuristic as for instance sequence the splits in the order of adsorbability or volatility. Denote the four components according to that order as 1, 2, 3 and 4. The associahedron K4 known also as MacLane pentagon condition is shown in Fig. 5.1 (MacLane 1971, Leinster 2004).

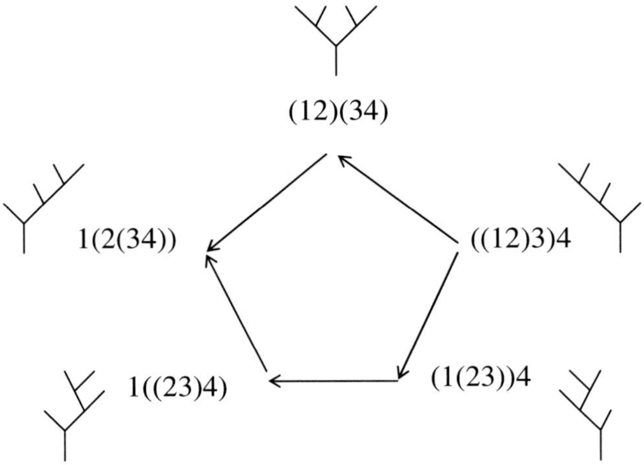

Fig. 5.1 Associahedron

Fig. 5.1 shows the possible separation schemas based on one property and outlines the relation between different schemas. For tensor product of four objects there are five ways to parenthesize it.

Observe that it is possible to build two isomorphisms from the sequence ((12)3)4 to 1(2(34)). The isomorphism is interpreted in the sense that the direct sequence 1(2(34)) is made equivalent to the reverse sequence ((12)3)4, if one retains the heavy phase instead of the light phase during the separation processes.

Table 5.1 contains the results of entropy calculus for different separation schemas as shown in Fig. 5.1.

To any component in a mixture we can associate a vector of properties and to separation schema shown in Fig. 5.1 we associate a matrix of properties.

Consider for instance the four vectors: 1=[1 1 1 1], 2=[1 1 1 0], 3=[1 0 0 1], 4=[1 0 0 0].

Applying the procedure of classification described in section 4.1 the separation schema

(1 2 3 4) → (1 2) (3 4) → (1) (2) (3) (4) results.

This is the first schema shown in Table 5.1.

The vector matrices have been selected to induce the desired separation schemas.

The entropy is calculated after different stabilization steps, for the similarity matrix (eq. 4.2). This stabilization step was denoted by n. The case n=0 corresponds to the entropy before any stabilization. For n=3 as number of stabilization steps all the similarity matrices associated to schemas shown in Table 5.1 became stabilized and invariant.

Observe that mirror symmetric schemas shown in Fig. 5.1 have the same informational entropy.

Table 5.1 Entropy calculus

Schema	Matrix	H (n=0)	H(n=3)
(12)(34)	1 1 1 1 1 1 1 0 1 0 0 1 1 0 0 0	7.021	6.989
1(2(34))	1 1 1 1 1 0 1 1 1 0 0 1 1 0 0 0	6.842	6.410
1((23)4)	1 1 1 1 1 0 1 1 1 0 1 0 1 0 0 0	6.794	6.410
(1(23))4	1 1 1 1 1 1 0 1 1 1 0 0 1 0 0 0	6.794	6.410
((12)3)4	1 1 1 1 1 1 1 0 1 1 0 0 1 0 0 0	6.842	6.410

Similarity matrix stabilization by max-min operations reduces the differences between the separation schemas and associated entropies.

A general study of coherence laws and higher operadic structures is due to Batanin (Batanin 2006). Fig. 5.2 shows the table of coherence laws illustrated by trees.

The table of coherence laws shown in Fig. 5.2 may be related to Baez and Dolan (1998) periodic table of categories.

The Baez and Dolan periodic table refers to k-tuply monoidal n-categories. A k-tuply monoidal n-category is an n-category in which objects can be multiplied in k ways, all of which interchange with each other up to isomorphism.

This implies that these k ways all end up being equivalent, but that the single resulting operation is more and more commutative as k increases. The stabilization hypothesis states that by the time we reach k=n+2, the multiplication has become maximally commutative. The stabilization hypothesis says that each column in the periodic table of n-categories stabilizes at a certain precise point.

Stability corresponds to symmetric monoidal categories in the periodic table of Baez and Dolan.

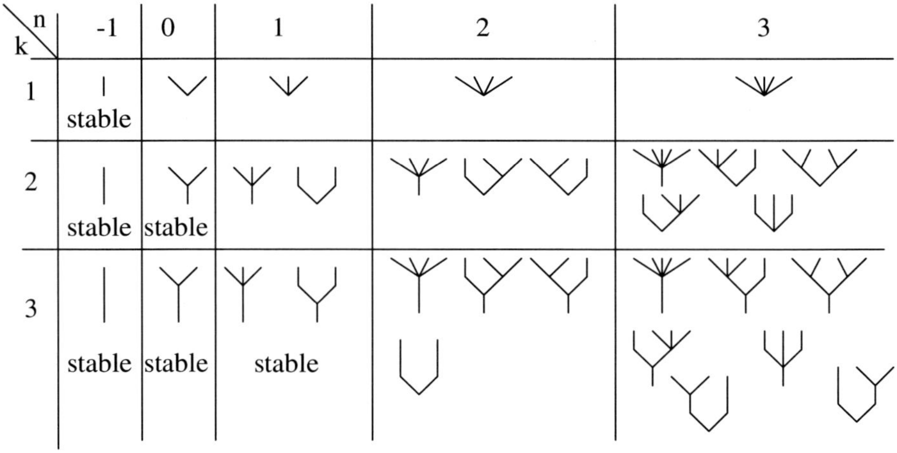

Fig. 5.2 Table of coherence laws

Fig. 5.2 shows a classification of the first few separation trees by n and k.

The coherence table due to Batanin was completed with a first column containing trees of width 1. This correspond to n = -1 (Baez and Shullman 2007).

This column describes a mixture that passes unsplit through k separation devices.

5.2 Symmetric Trees for Separation

A study of separation schemas may be based on categorical differentiation and Faà di Bruno construction (Cockett and Seely 2011).

This involves symmetric trees.

A symmetric tree of height $n \geq 0$ and of width $m > 0$, in variables $V = \{x_1, \ldots, x_m\}$, is defined inductively by:

- The only symmetric tree of height 0 has width 1 and is a variable y
- A symmetric tree of height $n \geq 1$, of width m, in the variables $\{x_1, \ldots, x_m\}$, is an expression $\bullet_r(t_1, \ldots, t_r)$ where each t_i is a symmetric tree of height n−1 in the variables V_i, where

$$\bigcup_{i=1}^{r} V_i = V.$$

The operation • is symmetric, so we are considering equivalence classes.

The inductive step involves splitting the variables into r disjoint non-empty subsets. The Stirling numbers of the second kind, S(n, r), are the number of ways of partitioning a set with n elements into r non-empty partitions.

We may regard the symmetric as algebraic expressions. In this view the operations at the nodes are symmetric, or commutative.

Consequently $\bullet_r(t_1, \ldots, t_r) = \bullet_r(t_{\sigma(1)}, \ldots, t_{\sigma(r)})$ for any permutation σ.

One can also regard these expressions as trees, in the graph theoretic sense, in which the leaves are uniquely labeled, as the root or by one of the variables, but in which no other node or edge is labeled.

Two symmetric trees are the same if graph theoretically they are isomorphic in a way which respects the leaf labeling.

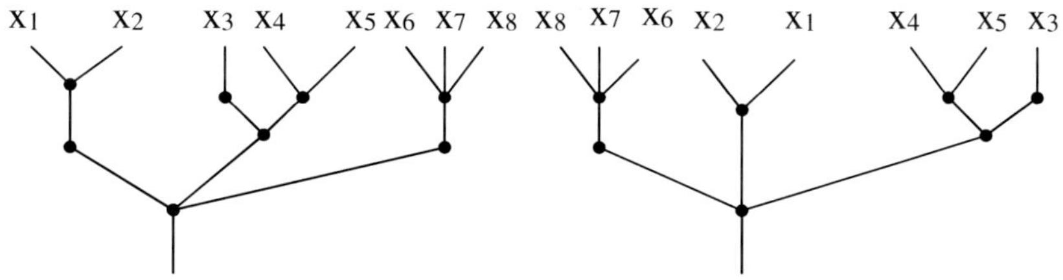

Fig. 5.3 Two representations of a symmetric tree

Fig. 5.3 shows two representations of a symmetric tree

A symmetric tree suggests a compact notation, representing the nodes as equivalence classes of the variables and then equivalence classes of these and so on. This allows us to represent the height 3 tree from Fig. 5.3 as: $\{\{\{x_1, x_2\}\}, \{\{x_3\}, \{x_4, x_5\}\}, \{\{x_6, x_7, x_8\}\}\}$.

Fig. 5.4 shows a classification of the first few symmetric trees by height and width.

If one wishes to generate all the symmetric trees of a given height and width one is presented by a combinatory problem as one must avoid generating trees which are already represented. There is a simple method for generating these trees.

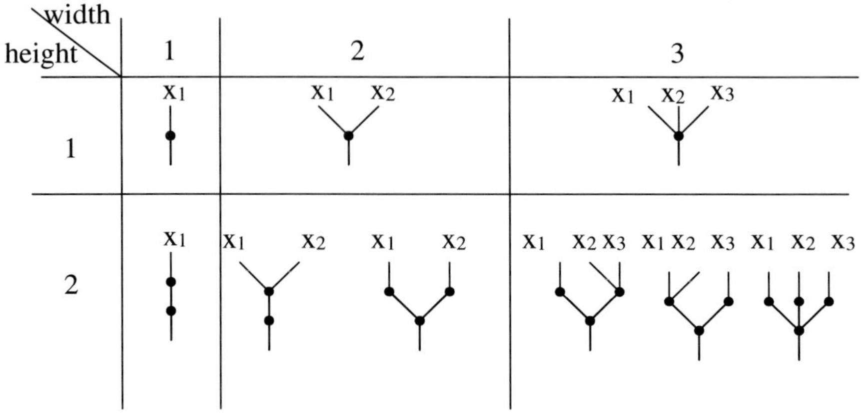

Fig. 5.4 Table of symmetric trees

The trees could be studied using a differential category construction.

The differential of a symmetric tree τ, of height n and width r produces a container of m trees of height n and width r + 1, where m is the number of nodes of τ . The new trees of the differential are produced by selecting a node and adding a branch to the new variable. The branch consists of a series of unary nodes applied to the new variable. These unary nodes are necessary in order to retain the uniform height of the tree.

For example the differential, introducing x_2, of the tree below is a pair of trees x_1 is shown in Fig. 5.5.

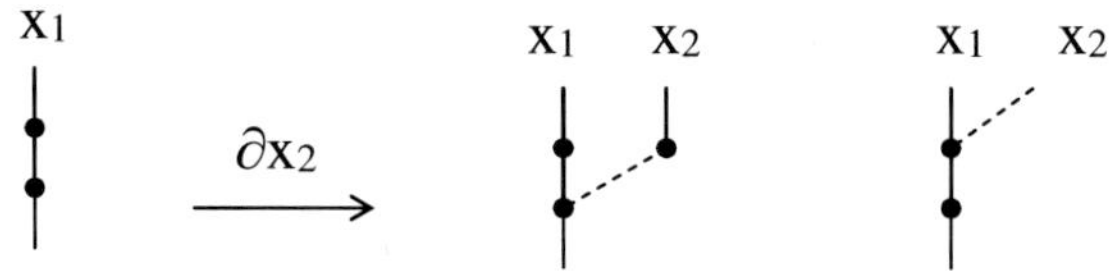

Fig. 5.5 Differential of tree

Every tree of height h and width d can be obtained as a member of the dth derivative of the unary tree of height h (Cockett and Selly 2011).

The construction is a rich potential source of differential algebras but also this structure is an integral part of what means to be differentiable.

The results shown in Fig. 5.4 and Fig. 5.2 suggest that the periodic table of categories could be related to differential categories and could be generated by differential models.

Cockett and Seely started by constructing a category Faà(*X*) with *X* a Cartesian additive category. A Cartesian category is a category with a terminal object and for any two objects in the category for instance A, B, the objects and mapping of the categorical product are in the category.

The objects of the Faà(X) are (A, A) pairs of object in X.

The morphisms are: f = (f_*, f_1, f_2,...) : (A, X)→ (B, Y) where: f_*:X→Y is a map in X

For r >0, f_r: AxA...xAxX → B is a symmetric form that is additive and symmetric in the first r arguments corresponding to A.

The composition is the Faà di Bruno convolution illustrated in Fig. 5.6.

Le τ define asymmetric tree of height 2 and width r, on variables $\{x_1,\ldots, x_r\}$

Then $(f\circ g)_\tau$: AxA.,...xA xX→C is defined by:

$$(f\circ g)_\tau = g_2(f_*(x), f_1(x_3, x), f_3(x_1,x_2,x_4,x)): AxAxAxAxX \to C \tag{5.1}$$

The convolution is defined by:

$$(fg)_n = \sum\nolimits_\tau (f,g)\circ\tau \tag{5.2}$$

Here $\tau \in T_2^n$ all the symmetric trees of height 2 and width n.

For any Cartesian left additive category X, Faà (X) is a Cartesian left additive category.

Cockett and Selly proved that Cartesian differential categories are exactly coalgebras of the Faà di Bruno comonad.

Cartesian differential categories have the so-called term logic calculus which make them much easier possible to work in (Blute et al. 2009). The aim was to make the term look like the standard notations for differential calculus. It was useful to develop the term logic for Cartesian differential categories not only so that the manipulation of maps is facilitated but also to illustrate the extent to which the intuitions from the ordinary calculus of differentiation are captured.

This means that it is possible to write differential equations that generate separation schemas, as the wave equation WE starts to do.

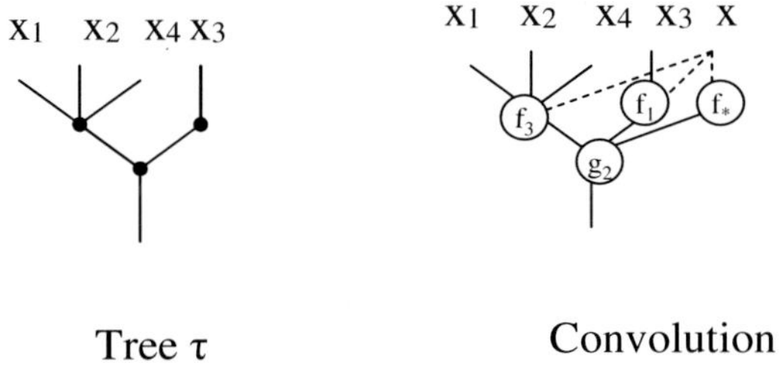

Fig. 5.6 Convolution

A symmetric tree may describe separation schemas in which the position of separation devices in the flowsheet is submitted to permutations.

In practice this can be accomplished with fixed separation devices if the access to different devices is driven by specially designed valves (Chin and Wang 2004).

Highly versatile separation systems are resulting.

It should be noted that the Faà di Bruno combinatorics was studied as Hopf algebra (Figueroa and Garcia-Bondia 2005).

5.3 Dual Graded Graphs for Separation

The theory of dual graded graph finds significant applications in the study of separation schemas.

It is known that separation schemas are based on difference in properties.

To every species in a mixture can be associated a vector the components of which take only two values "1" or "0" where "1" means the presence of a given property whereas "0" means its absence. For instance, "1" may correspond to high volatility, whereas "0" corresponds to a low volatility.

Vectors associated to different compounds are denoted by: $i= [i_1, i_2,..., i_k,...]$ where i_k are either "1" or "0".

Such vectors allow describing the trajectory of the compound in separation schemas.

For a given compound we may consider a separation schema that is a succession of devices. The schema evaluates level after level the compounds for different properties as volatility, absorbability, adsorbability and so on in an assumed significance order.

The compound is separated or not in the device focusing on that properties.

The vectors of properties are evaluated in the hierarchical schema or in schema resulting by some insertions of new devices, but not before the first one.

As an example we will consider the lifted binary trees and the Binword shown in Fig. 5.7.

The lifted binary tree (Fig. 5.7a) corresponds to the hierarchical separation schemas while the Binword (Fig. 5.7b) corresponds to separated schemas with insertions.

In the lifted binary tree y covers x if it is obtained by adjoining a single "0" or "1" to the end of x. On the other hand x is covered by y in Binword if it is obtained from y by removing a single letter, but not the first.

The resulting lifted binary tree and the Binword are dual (Fomin 1994).

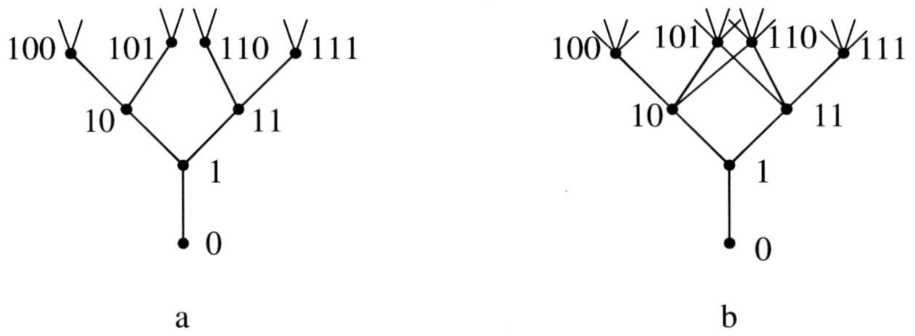

Fig. 5.7 Lifted binary tree and Binword

These are the graphs associated to the dual Hopf algebras NSym, noncommutative symmetric functions and Qsym, quasi-symmetric functions (Hivert et al. 2005, Nzeutchap 2006).

To clarify the functioning in duality we could consider the schema shown in Fig. 5.8.

The figure shows two graded graphs (G_1, G_2) corresponding to the two schemas. They have the same set of vertices and rank function.

The down operator D and the up operator U may be defined by the relations:

$$Uw = \sum_{v\,\mathrm{covers}\,w} m_1(w, v)v \quad ; \qquad Dz = \sum_{z\,\mathrm{covers}\,v} m_2(v, z)v \tag{5.3}$$

Here $m_i(x,y)$ denotes the multiplicity of (x, y) in G_i.

The thick arrows correspond to operators U and G_1 and the thin arrows correspond to operators D and G_2 in Fig. 5.8.

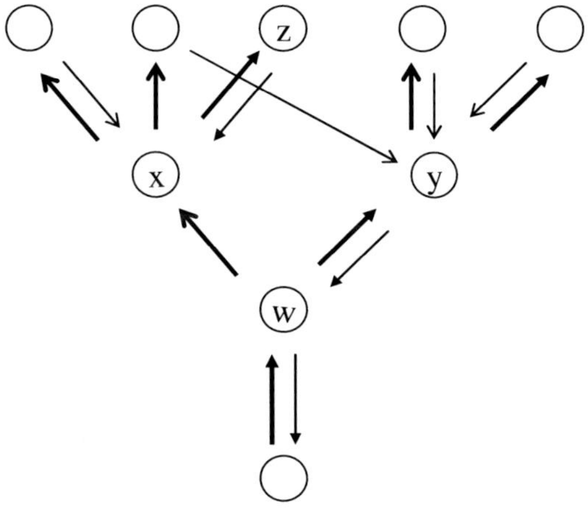

Fig. 5.8 Schema duality

The two graphs are different.

The dual graded graphs correspond to dual separation schemas.

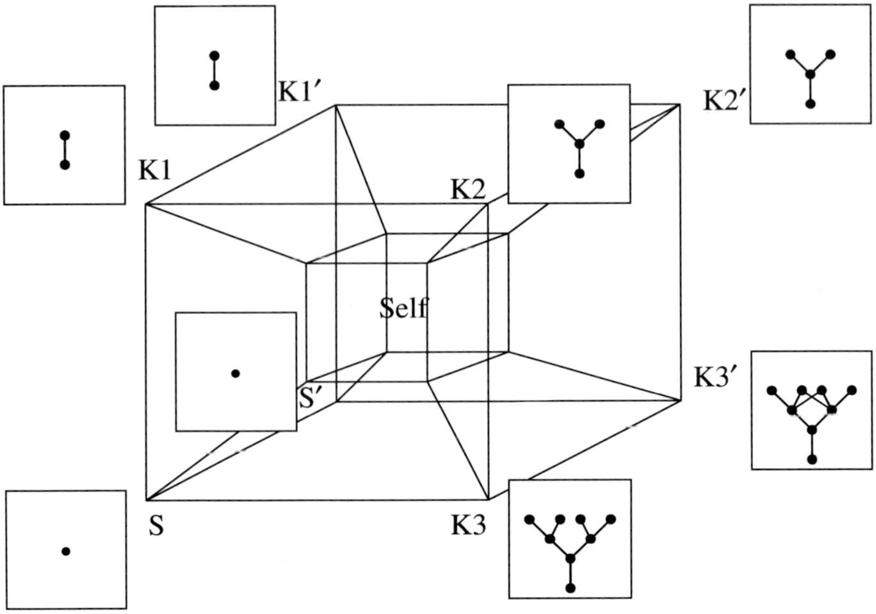

Fig. 5.9 Polytope for binary tree and Binword

Fig. 5.9 shows the lifted binary tree and the Binword construction. This illustrates the duality of the two separation schemas. The Self triggers the switching between the two schemas.

The binary trees of separation sequence S→K1→K2→K3 correspond to adjoining of new separation devices in hierarchical schemas.

The Binword trees sequence K3′→K2′→K1′→S′ corresponds to insertion of new separation devices in the existing hierarchical schemas, changing the order of separation properties except for the first one device and property.

It is the coupling of both schemas that allows process improvement as new separations, heat integration, reverse flow or periodic functioning and so on.

5.4 Tamari Lattices for Process Synthesis

The task of separating multicomponent mixtures into streams enriched in the respective constituents is commonly carried out in conventional separation columns arranged in series.

Due to restrictions for energy and cost, current research aims at alternative column arrangements that offer savings in energy and costs (Christiansen et al. 1997, Halvorsen and Skogestad 2011).

The schemas represented by trees are fundamental data structure in chemical engineering and they have been extensively studied.

One of the most common operations for restructuring a separation tree is described by the rotation operation shown in Fig. 5.10.

It refers to the rotation of the edge (x, y). Here R denotes right rotation and L denotes left rotation.

For binary separation columns the rotations correspond to the transition between indirect split and direct split (Fidkowski and Krolikowski 1987).

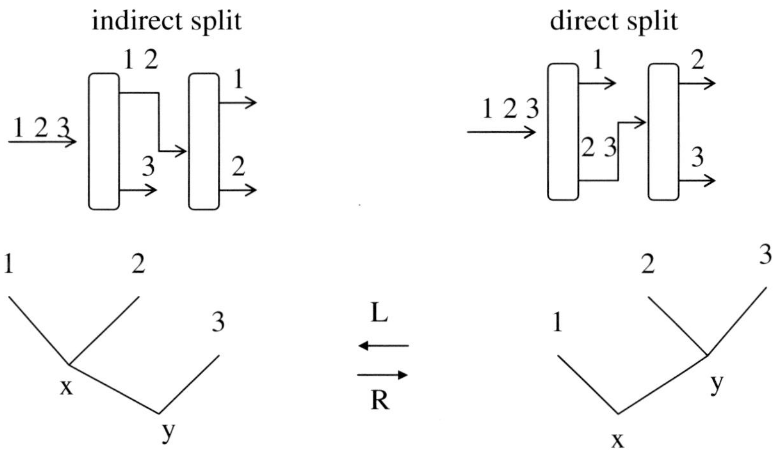

Fig. 5.10 Rotation operation

Such separation schemas may be organized in the poset of binary trees with n leaves, ordered by tree rotation operations. These are the so-called Tamari lattices.

A Tamari lattice is a partially ordered set in which the elements consist of different ways of grouping a sequence of objects into pairs using parentheses.

For instance, for a sequence of four compounds 1234, the five possible groupings are: ((12)3)4, (12)(34), (1(23))4, 1((23)4), and 1(2(34)). Each grouping describes a different order in which the compounds may be combined by a binary operation. In the Tamari lattice, one grouping is ordered before another if the second grouping may be obtained from the first by only rightward applications of the associative law (ab)c = a(bc).

Applying this law with $a = 1$, $b = 23$, and $c = 4$ gives the expansion $(1(23))4 = 1((23)4)$, so in the ordering of the Tamari lattice $(1(23))4 \leq 1((23)4)$ or, in other words, the sequence (1(23))4 is earlier than the sequence 1((23)4).

In this partial order, any two groupings g_1 and g_2 have a greatest common predecessor, the meet $g_1 \wedge g_2$, and a least common successor, the join $g_1 \vee g_2$.

The Hasse diagram of the Tamari lattice is isomorphic to the vertex-edge incidence graph of an associahedron. The number of elements in a Tamari lattice for a sequence of n + 1 objects is the nth Catalan number.

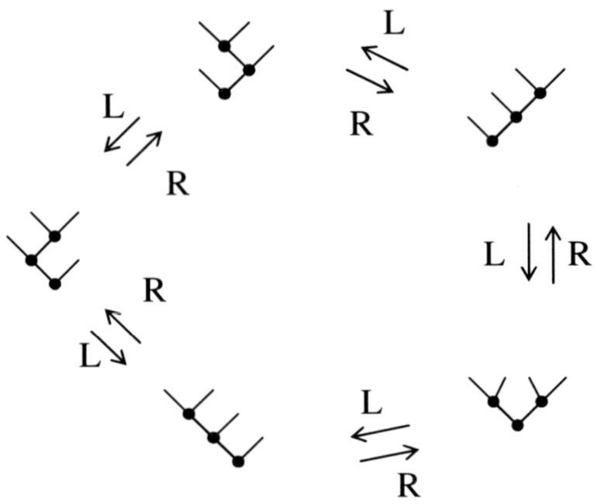

Fig. 5.11 Tamari lattice for 4-compounds separation trees

Fig. 5.11 shows the Tamary lattices with 3 nodes.

This lattice corresponds to 4-compounds separation trees.

Transition from a separation scheme to another is accomplished by rotation that is switching between direct and indirect sequencing.

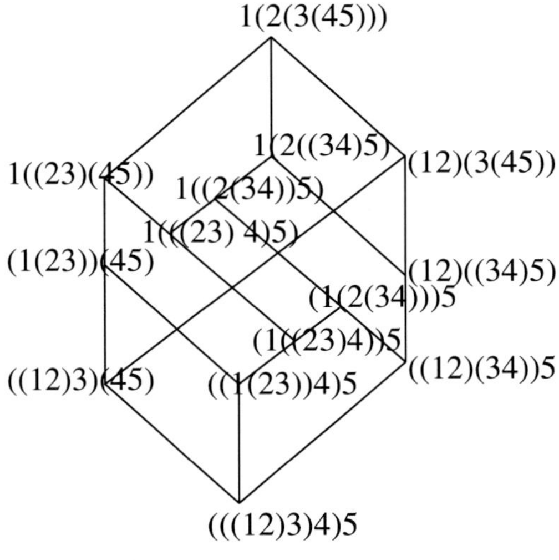

Fig. 5.12 Tamari lattice for 5-compounds separation trees

Fig. 5.12 shows the Tamari lattice associated to a 5-compounds separation.

It should be noted that the combinatorics of the Tamari lattice was studied in terms of Hopf algebras (Foissy 2009).

5.5 Cyclic Operations of Separation

Cycling operation methods are important in oil chemistry, in pharmaceutical and food industry, isotopes separation, hydrogen purification, desalinization, environment protection and so forth. Cyclic separation technologies such as pressure swing adsorption (PSA), temperature swing adsorption (TSA), vacuum swing adsorption (VSA), cyclic zone adsorption (CZA), simulated moving beds (SMB) chromatography, pressure swing reactor (PSR) and reverse flow reactor (RFR) are unsteady non-linear processes difficult to put into practice and to control.

Numerous cycling separation schemas, based mainly on intuition, have been reported in literature (Ruthven 1984, Yang 2003)). Well-known examples are the schemas involving 2 or 4-beds, and 2 or 4-step cycles. An example of SMB schema is shown in Fig. 5.13.

It is a basic configuration for cyclic operations.

The SMB consists of four columns or beds, #0, #1, #2 and #3, interconnected in a circular arrangement. The positions of feed, extract, desorbent and raffinate are changed cyclically in four steps corresponding to the four columns.

For multi-component, and multiple beds systems it is difficult to arrange the process scheduling, to elaborate and to run mathematical models, to adjust them by experiment.

The non-linear interactions of components and the interconnection of beds complicate the cycling operation schedule. As the separation complexity increases it becomes very difficult to formulate a feasible schedule much less an optimal one.

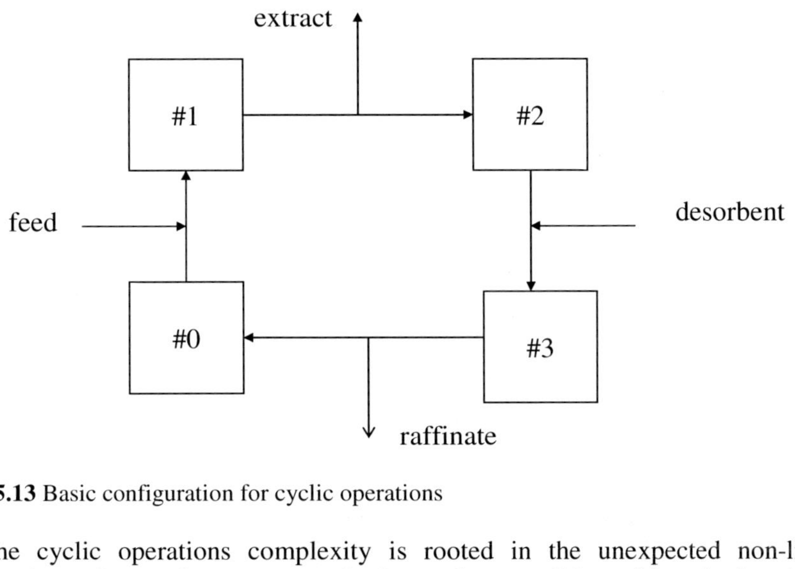

Fig. 5.13 Basic configuration for cyclic operations

The cyclic operations complexity is rooted in the unexpected non-linear interactions of several components, in the random condition of functioning, in the unavoidable uniqueness of each separation device, the incomplete knowledge of inputs and outputs, the variability in time of parameters, in modeling problems, and so forth.

The approach allowing operating cyclic separations in high complexity conditions is that of self-evolvable cycling separation systems. These are systems that can change autonomously both the schema as the dynamic behavior and are capable to control and to take advantage of the unexpected events of their environment in increasingly complex ways.

Self-evolvable devices are separation systems with emergent, and not pre-programmed, behavior.

Cyclic separation device with evolvability based on schema modification on self-configuring schemas and multi-scale schemas organized by self-similar replication at different conditioning levels may be characterized as operads (Iordache 2010).

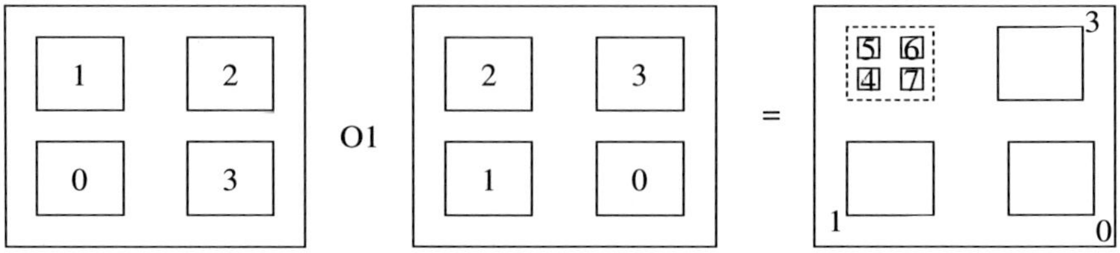

Fig. 5.14 Operad for cyclic operations

Fig. 5.14 shows the cyclic operation as an operad.

The operation O1 operates between the two squares. Each square is divided in four little squares.

The operation O1 consists in inserting in the fold 2 of the second square the structure of the first square. This is equivalent to a change of scale.

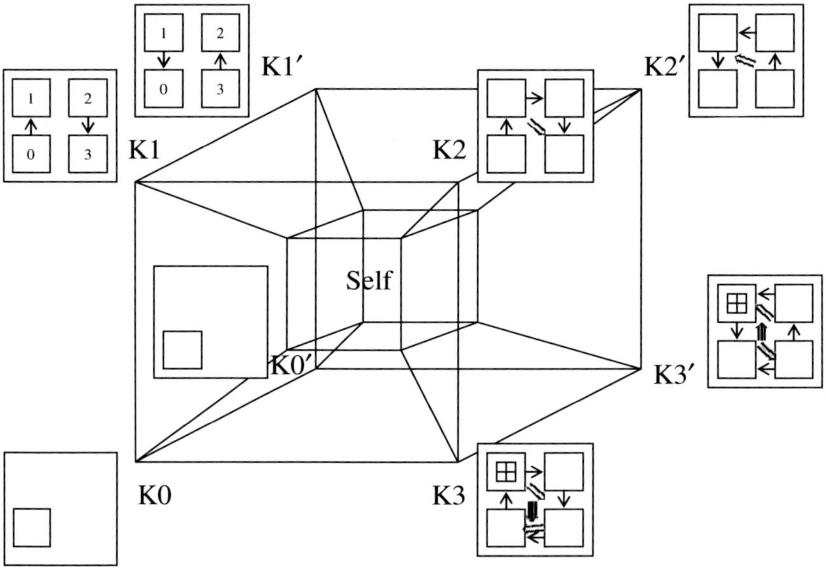

Fig. 5.15 Cyclic operations polytope

Fig. 5.15 shows an example of cyclic operations polytope.

Here after the direct way K0→K1→K2→K3 operations a shift from front face to back face illustrate a reverse order operation, K3′→ K2′→ K1′→ K0′.

The reverse order is obtained by reversing the arrows directions in a multi-scale schema (Iordache 2010). Double arrows correspond to 2-categories and triple arrows to 3-categories.

For an industrial system of separation schemas the two faces of the external cube may correspond to the cocurrent and countercurrent processing. The direct and reverse travel may correspond to activation and reactivation or other periodical regimes.

Making use of the direct way will give different results if compared to the reverse way. The swinging from direct to reverse cycling is beneficial since it allows online evaluations and new separations.

The swinging back and forth is mediated at the inner cube level, the Self.

Swinging is a tool for self-designing systems that can autonomously find separation solutions to highly complex and undefined separation problems. It allows the system self-evolvability.

5.6 Differential Model for Process Scheduling

The PSM approach is based on the differential model for symbolic aspects, the wave equation model, WE. This generates separation schemas and schedules. This model complements the dynamic mass transfer model. The wave model is formally similar to the first order wave micro-model used to describe separations (Rhee et. al. 1989) but the physical significance of parameters, the factors and the calculus rules are different.

The WE model allows simple description of flow sheets, characteristics and schedules. The connection with the theory of Latin squares and the designs of experiments was established and illustrated in the study of configurations with variable number of separation units or stages and in the study of operations coupling (Iordache 2010).

Consider now the convective part of the first order wave equation, WE:

$$\frac{\partial Y}{\partial T} \oplus V \otimes \frac{\partial Y}{\partial Z} = 0 \tag{5.4}$$

The initial condition is:

$$Y\,(Z, 0) = F\,(Z) = Z \tag{5.5}$$

This means that at T=0, the output Y of the separation schema at the distance Z in schema is exactly Z. The schema is one in which each separation level activates a new difference in properties allowing classification. In a separation schema since all the beds are operated with the same repeating sequence, each bed must be initiated with a one-step shift, along Z. This kind of initial condition ensures that the wave of the classification process is initiated and is going on.

The solution of the wave equation is:

$$Y=Z\oplus V\otimes T \tag{5.6}$$

The velocity V=1 in equation (5.6) will be considered in what follows.

This means that the characteristic solution is:

$$Y=Z\oplus T \tag{5.7}$$

The solutions of the model are in fact the addition operation for different C (m) algebraic frames.

• Example 1: Three Beds PSA Schema

Smith and Westerberg considered the following sequence of seven operations (Smith and Westerberg 1990):

O1-adsorption and production
O2-adsorption, production and production of purge gas for O5
O3-pressure equalization with low-pressure operation O6
O4-countercurrent depressurization
O5-countercurrent purge with gas from O2
O6-pressure equalization with high-pressure operation O3
O7-re-pressurize with feed gas

Table 5.2 shows the schema as presented by Smith and Westerberg:

Table 5.2 PSA configuration

<table>
<tr><td>#0</td><td colspan="3">O1</td><td>O2</td><td>O3</td><td>O4</td><td>O5</td><td>O6</td><td>O7</td></tr>
<tr><td>#1</td><td>O5</td><td>O6</td><td>O7</td><td colspan="3">O1</td><td>O2</td><td>O3</td><td>O4</td></tr>
<tr><td>#2</td><td>O2</td><td>O3</td><td>O4</td><td>O5</td><td>O6</td><td>O7</td><td colspan="3">O1</td></tr>
<tr><td>#3</td><td colspan="9">Reactivation</td></tr>
</table>

The three beds indexed by #0, #1 and #2 are the effectual and minimal ones.

The reactivation bed #3 was introduced since the adsorbents deactivates slowly.

To run separation in three beds the operation should be lumped in three classes, from the point of view of time spent. It is possible to lump the operations shown in Table 5.2 as follows: O1=0, (O2, O3, O4) =1, (O5, O6, O7) = 2. This is possible by ensuring that O1 has the same duration as the sequence (O2, O3, O4) or the sequence (O5, O6, O7). This is possible since due to process restrictions O2 and O5, O3 and O6 and implicitly O4 and O7 should have similar duration. For three beds #0, #1 and #2, three groups of operations denoted by 0, 1 and 2 will be considered.

Grouping the operations, Table 5.2 may be rewritten as Table 5.3.

Table 5.3 shows the three-bed schema for PSA.

Table 5.3 Three beds for PSA

Z\T	0	1	2
#0	0	1	2
#2	1	2	0
#1	2	0	1

• Example 2: Four-Bed Schedules

The four-bed schema is frequently encountered in cyclic operations.

Chiang presented the cyclic separation schema for PSA, shown in Table 5.4 (Chiang 1988).

Arrows show the flow direction. V-vacuum, E-equalization, R-re-pressurization.

A-adsorption, CD-counter-current depressurization, F-feed compression.

Table 5.4 Four-bed schedule for PSA

#0	V ↓	E ↓	R ↓	F ↑	A ↑		CD ↑	E ↑
#1	CD ↑	E ↑	V ↓	E ↓	R↓	F ↑	A ↑	
#2	A ↑		CD ↑	E ↑	V ↓	E ↓	R ↓	F ↑
#3	R ↓	F ↑	A ↑		CD ↑	E ↑	V ↓	E ↓

Considering the following lumping in blocks: (V, E) =0, (R, F) =1, A=2, (CD, E) =3 the Chiang schema reduces to the basic C (4) solution.

Notice that in the step 1 the flow direction may change after half period.

The same result may be obtained for the four beds Oxy-Rich process (Smith and Westerberg 1990, Fig. 2)

The notations to be considered in this case are: O1=0, O2=1, O3=2, O4=3

Table 5.5 shows the four-bed configuration.

Table 5.5 Four-bed configuration

Z\T	0	1	2	3
#0	0	1	2	3
#3	1	2	3	0
#2	2	3	0	1
#1	3	0	1	2

Similar results are obtained if we consider the cyclic schedule with four beds and seven operations (Smith and Westerberg 1990, Fig. 3).

The lumped operations to be considered are: O1=0, (O2, O3) =1, (O4, O5) =2, (O6, O7) =3.

Lumping imposes the time restrictions.

The same schema is resulting for simulated moving bed SMB chromatography (Ruthven and Ching 1989). Suppose that the SMB contains four zones of beds indexed by #0, #1, #2 and #3. The involved elements of the chromatography operation are: F-feed, E-extract (slow mover), D-desorbent (eluant), and R-raffinate (fast mover). They are denoted by "0", "1", "2" and "3" respectively. F and D are inlets, while R and E are outlets.

The functioning of beds at successive time steps is represented in Table 5.6.

Table 5.6 shows the SMB configuration.

Observe that the positions of the F, E, D, R are changed in the direction of circular flow at a regularly point in time. At any given time for any zone, only one of the valves corresponding to F, E, D, or R is open.

Table 5.6 SMB configuration

Z\T	0	1	2	3
#0	F	E	D	R
#1	E	D	R	F
#2	D	R	F	E
#3	R	F	E	D

• Example 3: Reconfigurable Separation Schemas

A reconfigurable array of modular micro-fluidic circuits inspired from cyclic operations of separation is presented in the following. In this case the circuit schema is associated to conditions, K, while the circuit of fluids and molecules through this schema is linked to the states, S. This complementarity is at the root of computing potentialities.

Fig 5.16 shows a reconfigurable array of cyclic separation modules.

The schema presented in Fig. 5.16 corresponds to an array of four-state modules similar to square tiles.

The four steps of any cell are F, E, D, and R. The FEDR rectangle or loop defines a single module.

The coupling of modules may be flexible. After each step, the F for the next module may be reached from E or from R of neighboring modules. The steps E or R should embed sensors and actuators.

The multiple cell stacked configuration is resulting if the tensor product interpretation in WE solution is a coproduct "$\cup$".

In quest for evolvability, transition from R→F connection type, to E→F connection type of two modules may be triggered by the presence of the answer of interest in steps R or E. The output of any module can be configured to be driven by its output or by signal arriving from a central programming unit for R or E gates.

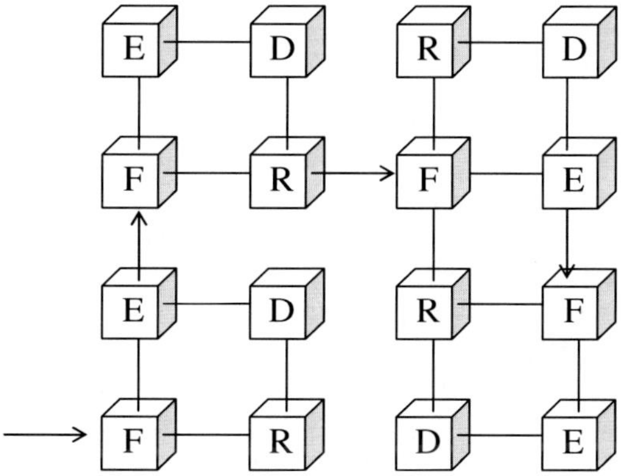

Fig. 5.16 Reconfigurable array of cyclic separation modules

The overall behavioral effects coming from the schema physical construction are implicitly taken into consideration. Only nearest–neighbor interconnections between modules were enabled in the schema from Fig 5.16.

Multiple levels of modules may be considered in 3-D space. The 3-D schema organized as Latin cubes based on orthogonal Latin offers interesting suggestions for high compactness of schemas.

For 2-D or 3-D schemas, recurrent connection paths through the schema by which a module output can indirectly affect its own input are possible.

An example is shown in Fig. 5.17.

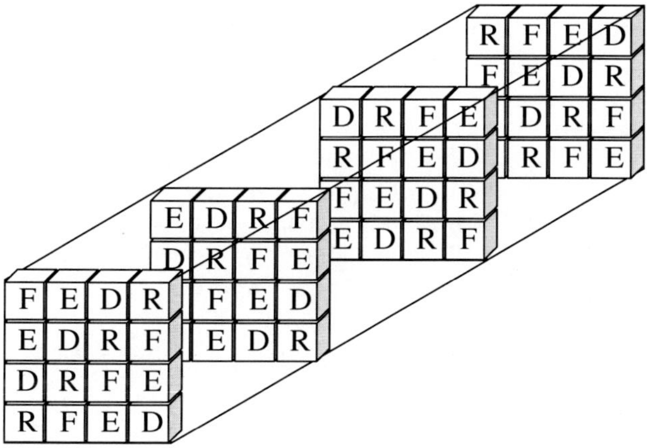

Fig. 5.17 Compact cyclic operations

Evolution is allowed to exploit the capability of the schema freely. Strongly interactive compounds may be forced to explore their space of possible schema and may create new unexpected patterns of interconnections and schemas.

5.7 Self-Evolvability for Circuits

Evolutionary circuits are circuits having the capability to change the preemptively embedded circuitry elements in order to keep on and to accomplish unprogrammed tasks. Evolutionary circuits make use of self-construction elements offered by the basic generic frame, and by the environment. Evolutionary circuits are expected to outline object-oriented behavior.

The *Pask's Ear* studied by Gordon Pask (Cariani 1989, 1993), the *evolved radio* described by Bird and Layzell, some developments of evolvable hardware may be regarded as evolutionary circuitry implementations (Bird and Layzell 2002).

One way to achieve circuit autonomy is to have sensors constructed by the system itself instead of sensors specified by the engineer. Cariani refers to Pask's system as a first example of such constructivist circuits. It is an electrochemical device consisting of a set of platinum electrodes in an aqueous ferrous sulfate/sulfuric acid solution. When current is fed through the electrodes, iron filaments tends to grow between the electrodes. If no or low current passes through a thread, it dissolves back into the acidic solutions. The threads that follow the path of maximum current develop the best. In the complex growth and decay of threads, the system mimics a living system that responds to rewards that is more current and penalty that is less current. The system itself is able to discover the most favorable forms for the condition, which may embed information concerning other factors of the environment such as magnetic fields, auditory vibrations, temperatures cycles. This circuit was trained to discriminate between 50 Hz and 100 Hz tones by rewarding structures whose conductivity co-varied in some way with an environmental perturbation. The Pask's device created a set of sensory distinctions that it did not previously have, proving that emergence of new relevance criteria and new circuits is possible in devices.

Following similar ideas, Bird and Layzell built an *evolved radio*. Like Pask's system the evolved radio determined the nature of its relation to environment and the knowledge of a part of the environment.

Bird and Layzell emphasized that novel sensors are constructed when the device itself rather than the experimenter determines which of the very large number of environmental perturbations act as useful stimuli.

Both of these devices, the *Pask's ear* and the evolved radio, show epistemic autonomy that is, they alter their relationship with the environment depending on whether a particular configuration generates rewarded behavior.

Fig. 5.18 represents a self-evolvable circuit as a polytope.

The notations are S-Environment, K1-Sensors, Perceptors, K2-Coordination, K3-Decision, Effectors.

The proposed architecture may be compared with Cariani cybernetic devices (Cariani 2008).

These devices, to varying degrees, are evolvable systems that continually modify their internal structure in response to experience. To the extent that an evolvable epistemic system constructs itself and determines the nature of its own informational transactions with its environment, that system achieves a degree of epistemic autonomy and self-evolvability relative to its surroundings.

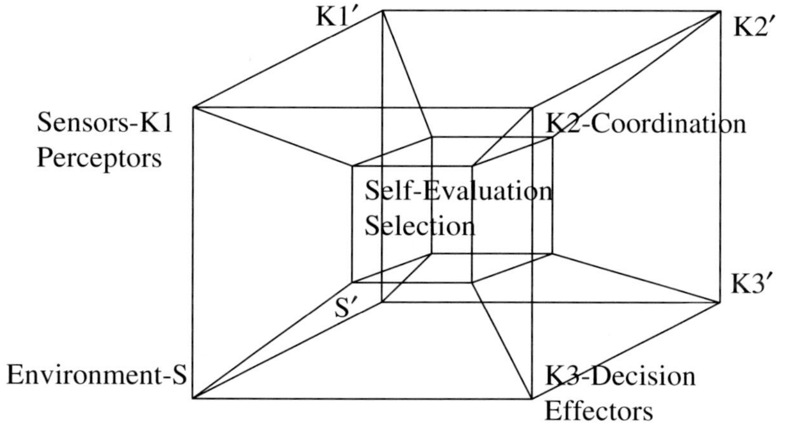

Fig. 5.18 Polytope for self-evolvable circuits

These devices consist of sensors, K1, and effectors, K3, coupled together by means of computational coordinative modules with well-defined internal symbolic states K2 (Fig. 5.18). These devices have an evaluative part, corresponding to the Self. This directs the construction and modification of the hardware that shows faculties similar to perception, cognition, evaluation, reward, and action. The Self module performs internal loops.

Their hardware includes sensors, effectors, and the internal computational mechanisms that mediate sensorimotor coordination by implementing particular percept-action mappings, corresponding to the external loops. The evaluative part contains memory, learning, and anticipatory mechanisms for measuring performance, changing percept-action mappings, and adaptively modifying internal structures to improve performance for the entire system.

A methodology able to distinguish between these functionalities and to determine when a new measurement, computation, or action is created was studied by Cariani (Cariani 2008). Such cybernetic systems can be described in terms of semiotic categories: syntactic, semantic, and pragmatic dimensions. Syntactics describes rule-governed linkages between signs that are implemented in computational, coordinative portions of devices. It corresponds to K2 in Fig. 5.18.

External semantics involves the relation of signs to the external world, that is, the causal linkages between internal symbolic states and the world that are mediated by sensors and effectors. It corresponds to K1 in Fig. 5.18.

Finally, pragmatics, corresponding to K3, involves the purposes for which signs are used: their relation to embedded goal states. Pragmatic relations are implemented by internal evaluation-reward mechanisms that adaptively steer or modify internal device linkages to better achieve embedded goals. Within such a framework one can envision devices with both mechanisms that swing between existing sets of possible internal states that is, combinatoric emergence or mechanisms that adaptively construct new hardware, that is, new sensors, effectors, internal states, capable of creating new functional primitives as creative emergence. In the syntactic realm, creative emergence produces new signs as symbols, or internal states. In the semantic realm, it produces new observables and actions that make new contingent linkages between internal states and the outer world. In pragmatic realm, it produces new evaluative criteria that is, new goals. Each function, that is, sensing, effecting, and coordinating can be either fixed, subject to combinatorial search, or capable of new primitive creation. In this schema, combinatoric creativity involves new combinations of pre-existing input and output states, sensors, effectors, and goals. Creative emergence requires going outside of the set of existing functionalities to modify material structures, hardware, in a manner that can create new states, new sensors and effectors, or new goals. To the degree that a system has control over its own structure and functions, it attains a degree of freedom relative to both its environment and its own history.

When a system can add to its own states and state transitions, as in a growing automaton, it achieves a degree of computational autonomy. When a system can construct its own sensors, it attains a degree of epistemic autonomy. When it can construct new effectors, it attains a greater autonomy of possible actions. Finally, when the system can construct its own set of evaluations and embedded goal states, it becomes self-directing.

This corresponds to the Self represented as the inner cube in Fig. 5.18.

We need to emphasize the role of the swinging behavior for the self-constructing devices.

Swinging is an important tool and goal for designing creative systems that can autonomously find solutions to highly complex and ill-defined construction problems.

Fig. 5.18 proposes that after the direct way of integration S→K1→K2→K3 we need to look at the reverse way of differentiation K3′→K2′→K1′→S′.

The Self module is able to correlate the two ways.

Making use of the developments of the direct way will result in different result for reverse way and this may correspond to the Self-capability to creatively change codes.

The boundaries where creativity grows and new information is created consist of synchronized direct and reverse ways, allowing self-evolvability.

Table 5.7 summarizes some of the differences between conventional circuits and the self-evolvable circuits.

Table 5.7 Comparison of conventional and self-evolvable circuits

Conventional circuits	Self- evolvable circuits
Single objective for any fabrication step	General classes of objectives
Defined based on previous learning	Undefined-open for learning, innovative, open-ended, creative
Top-down, linear	Top-down, bottom-up, multi-scale, cyclic
Aims for best solution, optimal	Makes workable, active, creative, self-evolvable
Looks for perfect elements	Accepts elements with small defects
Conventional design-detailed models	Generic design based on wave equation
Clear processing steps, complete data	Incomplete data and variable individual steps
Independent on previous designs	Use building blocks at hand, if useful
Insulate the elements, serial or sequential	Combine elements, distributed, parallel, swinging architecture
Builds	Builds, disbands, embeds, promote self-organization
Divide and conquer	Divide, integrates, centralizes
Functionality in different media	Sensitive to environment, multifunctional
Restricted, static	Less restricted, rich, self-dynamic
Isolate from medium protection	Medium, opportunistic, beneficial exploitation
Avoid variability, interactions, transitions	Accept, benefic use of variability, interactions
Reliable	Robust, resilient, multi-reliable
High maintenance	Anticipatory, dynamic resilience
Catastrophic degradation	Degradation in steps, hindered

5.8 Embedded Designs

Early use of biomolecules in information processing has been in the field of optical computing. This suggested as candidate for self-evolvable circuits, the bacterio-rhodopsin, BR, which can serve as computer switch (Birge 1995, Vsevolodov 1998).

BR has two useful properties for molecular-level calculation. It exhibits photochromic switching and shows photoelectric effect also.

The photo-cycle of BR, the sequence of structural changes induced by light-allows the storage of data in memory. Green, red and blue light induce structural changes of BR. Green light transforms BR in an intermediate denoted by “k” that relaxes to the “o”state. Red light transforms “o”state in “p” state that relaxes to “q” state. Blue light converts “q” state back to BR (Birge 1995). Any long-lasting states can be assigned to digital values making it possible to store information as a series of BR molecules in one or another state.

Discrete states as “0”, “1” and more than these are necessary for self-evolvable circuitry devices. With these identifications the BR substrate may be the source for the symbolic language such as pixels and strings.

The photoelectric effect is another BR property useful for self-evolvable circuit realization. Practical use of this property is exigent because it requires the preparation of BR films with highly oriented molecules. The possibility to interface BR electrically is the basis for several applications. The light of a specific wavelength range can be used to change the BR conformational state and the conformation change is accompanied by a color change that can be detected by optical means. It should be observed that the circuits are in this case, in part, of optical type.

A significant step in the development of the optoelectronic circuitry and computing was the study of all-light-modulated transmission mechanism of BR films. When a yellow beam and a blue beam illuminate the BR film, the two transmitted beams suppress mutually. Based on this mechanism, an all-optical operating device in which all 16 kinds of double-variable binary logic operations were implemented. The intensity of an incident yellow or blue beam acts as the input to the logic gate and the transmission bears the output of the gate. It is possible to turn this all-optical device into different states using different wavelengths and different intensity illuminations (Zhang et al. 2000).

The perspectives of a hybrid optoelectronic device based on BR molecules properties, in which conventional electronics is used to implement DOE analysis, are evaluated in the following.

Photo-cycle and photoelectric effects allow a direct writing DOE embedding in the BR based substrate. BR memorizing digits should be complemented by standard electronics able to perform the real-valued operations.

A computing cell with three BR molecules is retained here for illustration purposes.

Table 5.8 DOE associated to three molecules cell

Exp.	Molecule	Time	States
1	#0	0	G
2	#0	1	R
3	#0	2	B
4	#1	0	R
5	#1	1	B
6	#1	2	G
7	#2	0	B
8	#2	1	G
9	#2	2	R

Table 5.8 results due to Galois field, GF(3) calculations and corresponds to a 3x3 Latin square. The factors are the time steps 0, 1, 2, the molecules #0, #1, #2 and the states 0=G, 1=R, 2=B corresponding to the three colors green, red, blue able to induce transitions. The time is multiple of the same time step.

Standard DOE table may be developed by indicating the conditions associated to any element of the 3x3 Latin square.

The DOE selects the significant results and also the significant factors by standard ANOVA calculations done by an external computer. This is Fourier analysis over the real field, for the device functioning parameters.

Successive steps will continue the experiment in the direction of beneficial results. The new experiment means a new DOE based on GF(m) algebra calculation and the wave equation. Hardware may be achievable in 2-D or 3-D structures with concentric hierarchically located levels or planes. Light sources should be placed externally (Birge 1995). Fig. 5.19 shows a 3-D structure for Latin squares of order 3. Any face of the polyhedron contains Latin squares.

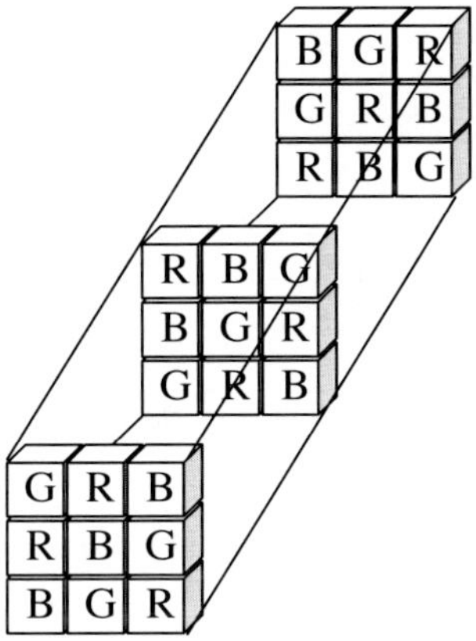

Fig. 5.19 Latin cube of order three

Based on special BR properties, new classes of self-evolvable circuits, embedding and evolving DOE became possible. The evolvability, for the proposed architectures is the expected result. As for the evolvable DOE structures, after the implementation of the DOE matrices it is required to perform at least two steps: factor evaluation, on columns in DOE, and experiment classification, on rows in DOE. It is necessary to define thresholds as degrees of acceptability for results. This helps to decide when to recognize a pattern to be classified, as new. Various areas throughout the chosen EC layers may be written and addressed simultaneously. It is conceivable to embed DOE matrices in any active areas with memory. Self-evolvable circuits would be built using in succession similar additive and subtractive steps as for printed circuits and integrated circuits fabrication. DOE matrices play the role of masks in circuits' fabrication. These self-evolvable circuits should be able to drive the input signal and to decode the signal in a manner similar to logical thinking processes. As a difference, if compared to conventional circuits, this kind of self-evolvable circuits will be continuously formed and erased, allowing the operation to be in succession in direct way and reverse way. The parallel search may be organized to achieve amplification, resonance and coherency. The self-evolvable circuits work associatively as well as serially. By parallel processing the experiments would be performed at once, and the recorded results can be presented simultaneously to the Self. The self-evolvable circuits should be able to record data from different areas to analyze and to give rise to a decision. This means that it need to have monitoring functions, that is sensors, and executive functions, that is actuators, since the long term technological challenge is to get results by self-evolvable circuits, independent of any external analyst or operator. The circuits should be a system that confronts the environment having the ability to adapt autonomously. New environmental conditions for circuits may be materialized by a new row in the existing, embodied, component DOE matrices. This is the discrete symbolic step of the self-evolvable circuits. Then follows the step in which real field values are associated to discrete DOE. This real-valued step goes after data expansion and precedes data compression. With a learned degree of acceptability the sensor information goes backward and is classified. In this way the material embodiment may regenerate the symbolic description represented by DOE.

5.9 Electrochemical Filaments Circuits

Based on electrochemical filaments development, a new type of evolutionary circuits, the electrochemical filaments circuits, ECFC became possible.

ECFC construction starts with a generic framework representing the elements of the set of conditions K.

The K-framework elements may be that generated by wave equation, WE. The process in K generates successive K-frames that is, K0, K1, K2, and K3 at different conditioning levels.

The generic circuitry represented by K-frames, is completed by additional circuitry, traces, dendrites, filaments, and supplementary matter, corrosion or degradation products. The processing for these additional circuits is an S-process. S denotes the physical circuit based on filaments, threads, or micro-channels for fluids allowing the electrical contact or interaction. The K-steps and the real environment S-steps have complementary contributions in circuit building. ECFC are expected to be at least partially self-evolvable. The self-evolvability includes the capability in building, assembly, modifying, organizing, repairing and destroying. As a difference, if compared to adaptive and self-adaptive devices based mainly on feedback, ECFC make use of the preemptively embedded K frames. The appropriate K designs and the selective addition and the subtraction of appropriate elements from environment are the processes allowing both self-functionality and self-evolutionary behavior.

The basic elements of ECFC technology are the K-valued generic framework, linked to classes of tasks, the environment media for self-construction in non-stationary or oscillatory fields and the self-learning capability by exposure to environmental complexity and to variable tasks.

The ECFC that results by coupling the electrochemical filaments, ECF of different orders m, ECFm, over pre-existing K-frame, K0, K1, K2, and K3 is considered here. The circuit may be described using the categorical tensor" $*$ " that links different levels in circuitry: $ECFC = K0 * K1 * K2 * K3 * ECF0 * ECF2 * ECF3$.

The tensorial product" $*$ "may be the categorical product "$\times$ ", the coproduct "$\cup$" and so on according to the categorical level.

The K-framework should be a quasi-complete printed circuit, with several opens and closures. The swinging between opens and closures allow the ECFC versatility and multiple potentialities. The environment is able to fill the opens sequentially in a way that ensure functionality. Potential geometrical variants and architectures for ECFC are: dots, cells, hexagons, triangles, squares, circles arrays, circular crowns, dyadic structure, labyrinths and mazes, self-similar nested structures, tiles, fractals and polytopes.

It was established that the wave equation, WE, is able to generate fractal structures making use of categorical product "$\times$ ". For example, Hadamard-Sylvester matrices reduce to Sierpinski triangles if only the "1"s are considered while the "-1"'s or with other notations the "0"s are neglected since they break the circuit (Iordache 2010).

The swing from categorical product to categorical coproduct determines the size and the shape of the circuit. The swing is determined by the oscillatory fields that accompanies the ECFC development.

The materials should offer opportunities for wet chemistry and for solid physics to play significantly. ECFC's make use of composites and multi-phase media. The materials should be as rich as possible in structural possibilities, for example in phase transitions, on the edge of chaos, in non-linear regimes. Interesting options are the existing self-adaptive or smart materials that allow phase transition, such as the piezoelectric, thermoelectric, electrorheological, electro-active polymers and so on.

Laminate known as filaments non-resistant as polyester rigid woven glass, paper phenol, or specially contaminated laminates represents valid opportunities since they allow the electrochemical filament fast formation.

Possible K-frames conductor lines make use of materials like Cu, Ag, Sn, Zn, Al, Mg and Fe. Metallic inorganic salts for conduction may be: sulfates, chlorides or nitrates of Fe, Cu, Ag, Sn, Pd, Pt, Zn, Al, Mg and catalysts. Metallic oxides may be useful as potential dielectrics. Damaged or fatigued printed or integrated circuits represent new potentialities for proactive behavior.

ECFC should be processed in the environment that is in real field conditions in which the circuit should be functional such as:

- Mechanical vibrations
- Temperature, relative-humidity, bias with direct current, alternative current or pulse plating of variable frequency
- Light, radiation
- Cyclical operation of various types
- Superposed oscillatory fields

These kinds of fields are the usual field of stresses for reliability tests. This suggests that evolutionary circuits may result from some over-tested circuits still able to show new capabilities.

The ECFC would be a circuit useful and stable in its building conditions. For any new level another frequency domain of oscillatory field is associated. As much as the oscillatory field still exists, the new level would be developed. If the structured filaments structures were located in a specific field, the resulting structure would be able to recognize the patterns of that field. Learning and removal of information is possible if any filament may continually be formed broken and regenerated. Training to discriminate signals may be accomplished with the help of wave equation WE solutions.

It should be noted that self-organization of similar systems of interacting particles or micro-robots was studied using graph grammars theory (Klavins 2007).

5.10 Dendritic Growth

The dendritic or filaments growth was studied using operads of Young diagrams (Forcey et al. 2007).

Examples of n-fold monoidal categories include ordered sets with n different binary operations. For each pair of operations an inequality expresses the interchange.

The additions are vertical and horizontal stacking, and the multiplications may be two ways of packing one Young diagram into another based respectively on stacking first horizontally and then vertically, and inversely.

The n-fold monoidal categories generalize braided and symmetric categories while retaining precisely enough structure to support operads. The category of n-fold operads inherits the iterated monoidal structure. The sequences that are minimal operads in the totally ordered categories introduced, and how these sequences grow represents the object of the study (Forcey et al. 2007).

It was observed that the growth rate of physical filaments or dendrites oscillates in a way directly comparable to that of the operads.

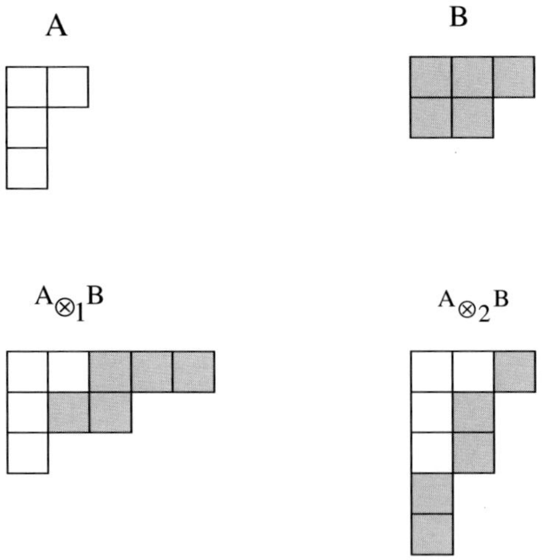

Fig. 5.20 Young diagrams for planar dendritic growth

Fig. 5.20 shows a Young diagram for planar dendritic growth. A and B are dendrites.

Here $\otimes_1$ denotes the horizontal stacking while $\otimes_2$ denotes the vertical stacking.

Fig. 5.21 shows a 3D-Young diagram for spatial dendritic growth.

Here $\otimes_1$ denotes the z-axis stacking that is the vertical concatenation of matrices followed by sorting the new longer columns.

The product $\otimes_2$ denotes the y-axis stacking that is the horizontal concatenation of matrices followed by sorting the new longer rows.

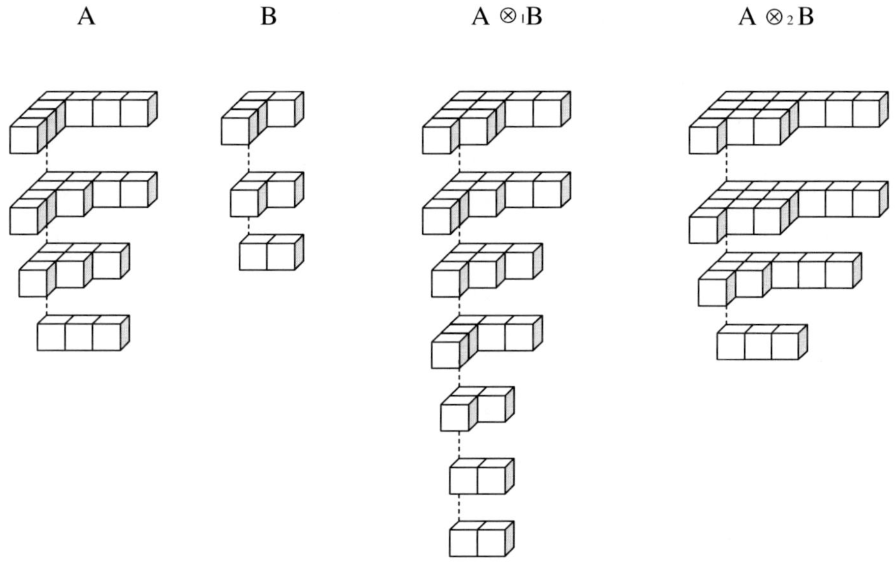

Fig. 5.21 Young diagrams for spatial dendritic growth

The corresponding matrices are:

$$A=\begin{bmatrix} 4 & 3 & 1 & 1 \\ 4 & 2 & 1 & 1 \\ 3 & 2 & 1 & \\ 1 & 1 & 1 & \end{bmatrix} \qquad B=\begin{bmatrix} 3 & 1 \\ 2 & 1 \\ 1 & 1 \end{bmatrix} \tag{5.8}$$

$$A \otimes_1 B = \begin{bmatrix} 4 & 3 & 1 & 1 \\ 4 & 2 & 1 & 1 \\ 3 & 2 & 1 & \\ 3 & 1 & 1 & \\ 2 & 1 & & \\ 1 & 1 & & \\ 1 & 1 & & \end{bmatrix} \qquad A \otimes_2 B = \begin{bmatrix} 4 & 3 & 3 & 1 & 1 & 1 \\ 4 & 2 & 2 & 1 & 1 & 1 \\ 3 & 2 & 1 & 1 & 1 & \\ 1 & 1 & 1 & & & \end{bmatrix} \qquad (5.9)$$

Only the non-zero entries of the matrices are shown.

The model was functional to describe the measurements of certain crystals formed in solutions. The fact that at certain temperatures the usual regular increase in size of the crystal became a pulsating, rhythmic growth find natural explanation with this type of models (Ferreiro et al. 2002).

The dendritic trees may be studied using the lattice of binary tree and the bracket tree (Fig. 5.22).

This is another example of duality between graded graphs. The vertices of rank n are given by the binary tree with n nodes.

In the lattice of binary trees, a tree covers exactly those trees obtained from it by removing a single leaf (Fig. 5.22a).

In the bracket tree, a tree covers a tree obtained by deleting and contracting the edge, if any, below the leftmost node (Fig. 5.22b).

This pair of dual graded graphs is associated to the Hopf algebra of planar binary trees and its dual (Loday and Ronco 1998).

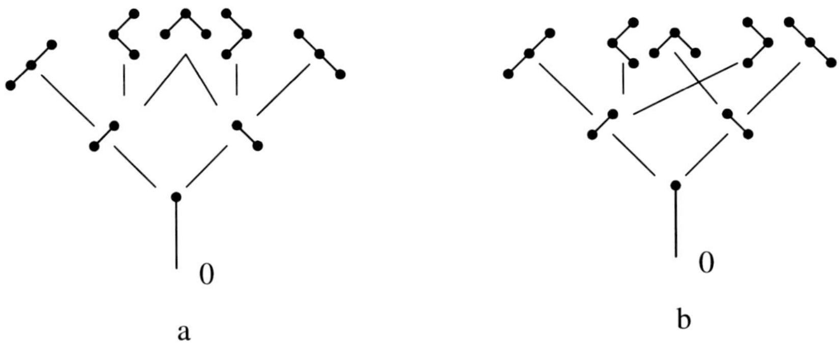

Fig. 5.22 Lattices for binary tree and bracket tree

The combinatorics of rooted tree systems similar to dendritic growth was studied using Hopf algebra and operads (Chapoton and Livernet 2001).

Fig. 5.23 shows the binary tree and the bracket tree construction. This illustrates the duality of the two growth schemas. The Self should trigger the switching between the two schemas.

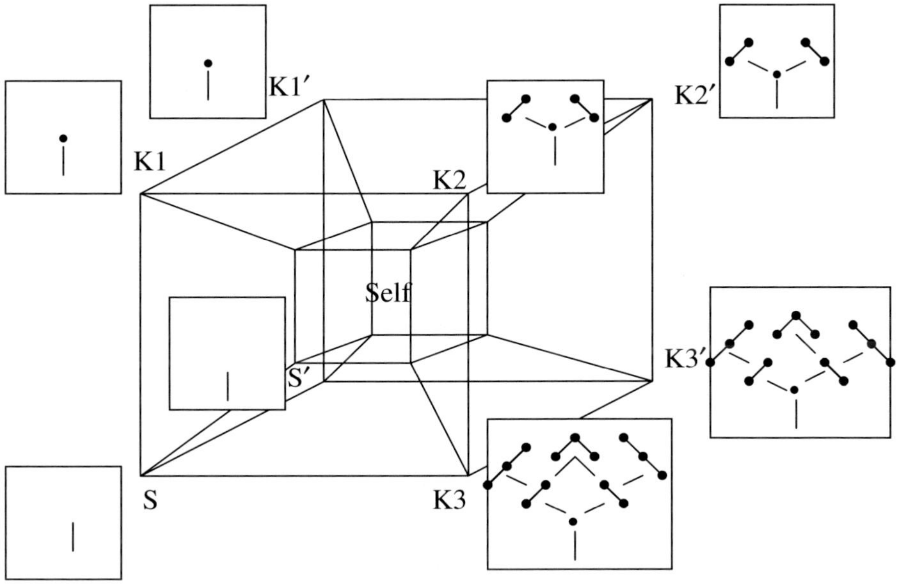

Fig. 5.23 Polytope for binary tree and bracket tree

Fig. 5.24 shows operads for dendritic growth.

Here the central module indicates the way to put together, while the surrounding modules contain the things to put together.

The dendrite development for ECFC suggests that their capabilities, are at the level of 1-categories and 2-categories. For 2-categories the pentagon relation is valid.

ECFC is able to disconnect a dendrite and reconnect in another position.

This corresponds to 2-categories. As operadic level this corresponds to the associahedron K_4.

For 3-categories the pentagon of pentagons or the associahedron K_5 should be considered. This need a spatial view that may exceeds the ECFC capabilities.

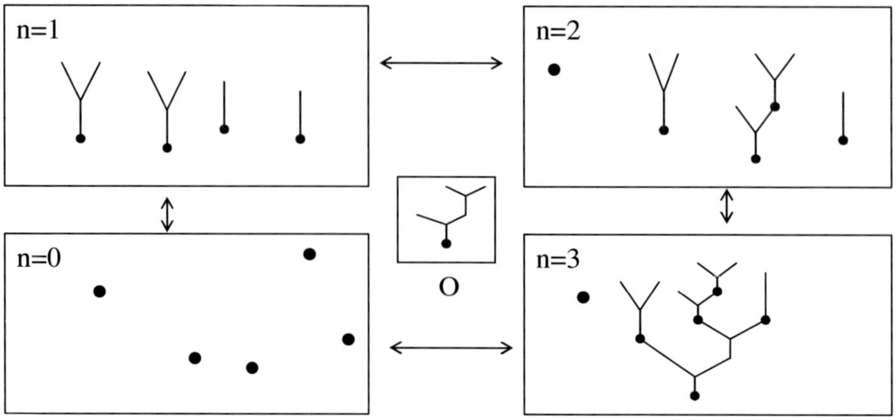

Fig. 5.24 Operads for dendritic growth

5.11 Self-Evolvable Antennas

Antennas may be generated as fractal polytopes (Colthurst 1996, Pearse 2007)

Fractal polytopes are fractals formed by repeated replacing a polytope with smaller polytopes at its vertices.

This family of fractals contains many fractal constructions such as the Sierpinski gasket and the Cantor set.

For every regular polytope, a just-touching regular polytope may be obtained.

Consider a polytope P, the n-dimensional generalization of a polygon or polyhedra.

The fractal polytope F (P,R), 0<R<1, with N vertices, is defined as the limit of the construction which takes P and replace it with N smaller polytopes, each with edge length R times that of P and placed at P's vertices (Fig. 5.25)

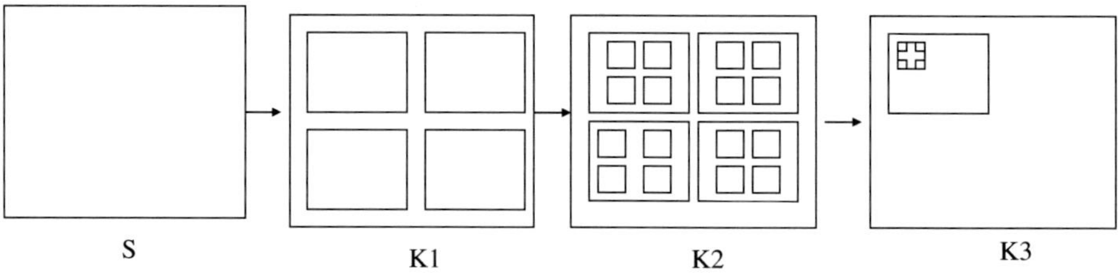

Fig. 5.25 Fractal antenna development stages

By construction F (P, R) is easily seen to have similarity dimension log N/log(1/R) and thus deserves the name fractal.

This construction may be defined as an iterated function system (Barnsley 1993).

Fig 5.24 emphasizes the fractal polytope in relation to different stages S, K1, K2 and K3.

Other choices of the stages are illustrated is shown in Fig. 5.26.

The notations are: S-Antenna, K1-RF Receiver, K2-Evaluation Software, and K3-Control Interface.

This refers to antennas as described by Linden (Linden 2002).

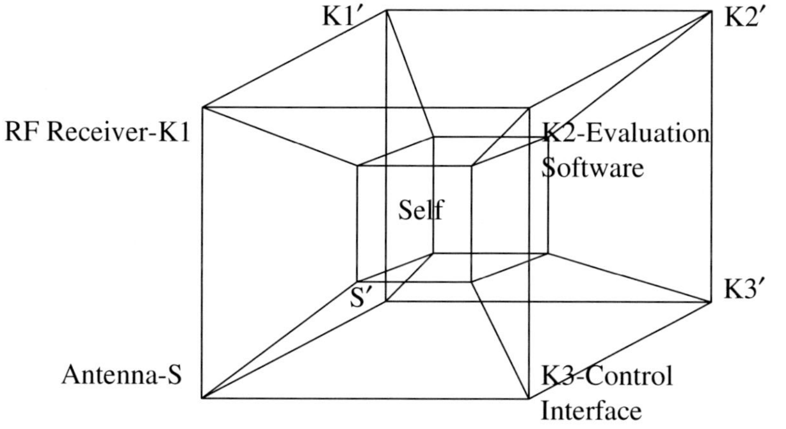

Fig. 5.26 Polytope for self-reconfigurable antenna

Such antennas may self-adapt to change in configuration and orientation as well as to damage.

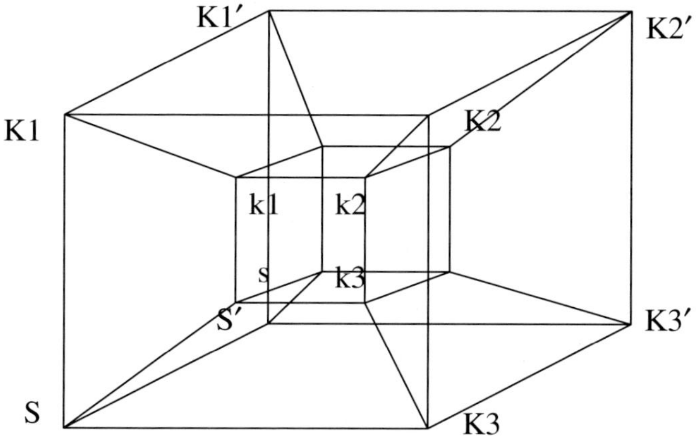

Fig. 5.27 Polytope for self-evolvable antennas

There exists a limitation of the number of levels for antenna development.

However, from an electromagnetic-wave point of view this limitation is not outstanding: from a certain iteration the electromagnetic waves are unable to resolve much smaller intricacies than a wavelength, and using highly iterated devices is not useful due to the high ohmic losses and the large stored energy in the surroundings of the pre-fractal (Gianvittorio et al. 2001, Gianvittorio 2003).

A self-evolvable antenna that varies in time should be considered.

Romeu and Blanch discussed the case of Hilbert curves (Romeu and Blanch 2002).

A polytope is shown in Fig. 5.27. In this case K1 is 1-D Hilbert curves, K2 is 2-D Hilbert curves, and K3 is 3-D Hilbert curves. The central cube corresponding to the Self indicates the way to put together such modules. In this case the sub-modules denoted by s, k1, k2 and k3 are in some way associated to S, K1, K2 and K3 modules. It is a smart antenna having an automatic evolution. Signal reception is ensured by cooperation of all the levels.

Fig. 5.26 and Fig. 5.27 propose that after the direct way of integration or convergence S→K1→K2→K3 we need to look at the reverse way of differentiation or divergence K3′→K2′→K1′→S′.

The Self module is able to mediate swinging and to correlate the two–ways.

Tendencies to converge for signals should coexist with tendencies to diverge and it is the rhythm and blend of both that matters and allows self-evolvability.

References

Baez, J., Dolan, J.: Categorification. In: Getzler, E., Kapranov, M. (eds.) Higher Category Theory. Contemp Math. Amer., vol. 230, pp. 1–36, Math. Soc. (1998)

Baez, J., Shulman, M.: Lectures on n-categories and cohomology. arXiv:math/0608420v2 (2007)

Barnsley, M.F.: Fractals Everywhere. Academic Press, New York (1993)

Batanin, M.: Confguration spaces from combinatorial, topological and categorical perspectives (2006), http://www.maths.mq.edu.au/street/BatanAustMSMq.pdf

Bird, J., Layzell, P.: The Evolved Radio and its Implications for Modeling the Evolution of Novel Sensors. In: Proceedings of the Congress on Evolutionary Computation, CEC 2002, pp. 1836–1841 (2002)

Birge, R.: Protein-based computers. Sci. Am. 272, 90–95 (1995)

Blute, R.F., Cockett, J.R.B., Seely, R.A.G.: Cartesian Differential Categories. Theories and Applications of Categories 22, 622–672 (2009)

Cariani, P.: On the Design of Devices with Emergent Semantic Functions. PhD. Thesis. Binghamton University, NY (1989)

Cariani, P.: To evolve an ear: epistemological implications of Gordon Pask's electrochemical devices. Systems Research 10, 19–33 (1993)

Cariani, P.: Design Strategies for Open-Ended Evolution. In: Bullock, S., Noble, J., Watson, R., Bedau, M.A. (eds.) Artificial Life XI, Proceedings of the Eleventh International Conference on the Simulation and Synthesis of Living Systems, pp. 94–101. MIT Press, Cambridge (2008)

Chapoton, F., Livernet, M.: Pre-Lie algebras and the rooted trees operad. Int. Math. Res. Not., 395–408 (2001)

Chiang, A.S.T.: Arithmetic of PSA process scheduling. AIChE J. 34(11), 1910–1912 (1988)

Chin, C.Y., Wang, L.N.H.: Simulated Moving Bed Equipment Designs. Separation and Purification Reviews 33(2), 77–155 (2004)

Christiansen, A., Skogestad, S., Lien, K.: Complex distillation arrangements: Extending the Petlyuk ideas. Comput. Chem. Eng. 21, S237–S242 (1997)

Cockett, J.R.B., Seely, R.A.G.: The Faà di Bruno Construction. Theory and Application of Categories 25, 394–425 (2011)

Colthurst, T.: Fractal Polytopes (1996) (preprint)

Ferreiro, V., Douglas, J.F., Warren, J., Karim, A.: Growth pulsations in symmetric dendritic crystallization in thin polymer blend films. Phys. Rev. E 65, 051606, 1–16 (2002)

Fidkowski, Z., Krolikowski, L.: Minimum Energy Requirements of Thermally Coupled Distillation Systems. AIChE J. 33(4), 643–653 (1987)

Figueroa, H., Gracia-Bondia, J.M.: Combinatorial Hopf algebras in quantum field theory I. Reviews in Mathematical Physics 17, 881–976 (2005)

Foissy, L.: The infinitesimal Hopf algebra and the poset of planar forests. J. Algebraic Comb. 30, 277–309 (2009)

Forcey, S., Siehler, J., Seth Sowers, E.: Operads in iterated monoidal categories. Journal of Homotopy and Related Structures 2, 1–43 (2007)

Gianvittorio, J.P.: Fractals, MEMS and FSS Electromagnetic Devices: Miniaturization and Multiple Resonances. PhD. Thesis. UCLA Electrical Engineering Department (August 2003)

Gianvittorio, J.P., Rahmat-Samii, Y., Romeu, J.: Fractal FSS: various self-similar geometries used for dual-band and dual-polarized FSS. In: Proc. IEEE Antennas and Propagation Soc. Int. Symp., vol. 3, pp. 640–643 (2001)

Halvorsen, I.J., Skogestad, S.: Energy efficient distillation. Journal of Natural Gas Science and Engng. 3(4), 571–580 (2011)

Helfferich, F.G., Klein, G.: Multicomponent Chromatography. Marcel Dekker, NY (1970)

Hivert, F., Novelli, J.-C., Thibon, J.-Y.: The algebra of binary search trees. Theoretical Computer Science 339, 129–165 (2005)

Iordache, O.: Polystochastic Models for Complexity. Springer, Heidelberg (2010)

Klavins, E.: Programmable self-assembly. Control Systems Magazine, IEEE 27(4), 43–56 (2007)

Leinster, T.: Higher Operads, Higher Categories. Cambridge University Press (2004)

Linden, D.S.: Optimizing Signal Strength In-Situ Using an Evolvable Antenna System. In: 2002 NASA/DoD Conference on Evolvable Hardware, Alexandria Virginia, USA, July 15-18, pp. 147–151. IEEE Computer Society (2002)

Loday, J.L., Ronco, M.O.: Hopf algebra and the planar binary trees. Adv. Math. 139(2), 293–309 (1998)

MacLane, S.: Categories for the Working Mathematician. Springer, New York (1971)

Nzeutchap, J.: Dual graded graphs and Fomin's r-correspondences associated to the Hopf algebras of planar binary trees, quasi-symmetric functions and noncommutative symmetric functions. In: International Conference on Formal Power Series and Algebraic Combinatorics, FPSAC, San Diego (2006)

Pearse, E.P.J.: Canonical self-similar tilings by iterated function systems. Indiana Univ. Math. J. 56, 3151–3170 (2007)

Rhee, H.K., Aris, R., Amundson, N.R.: First Order Partial Differential Equations II. Theory and applications of hyperbolic systems of quasilinear equations. Prentice-Hall, Englewood Cliffs (1989)

Romeu, J., Blanch, S.: A three-dimensional Hilbert antenna. In: Proc. IEEE Antennas and Propagation Society Int. Symp., San Antonio, TX, vol. 4, pp. 550–553 (June 2002)

Ruthven, D.M.: Principles of adsorption and adsorption processes. John Wiley, New York (1984)

Ruthven, D.M., Ching, C.B.: Counter current and simulated counter current adsorption separation processes. Chem. Engng. Sci. 44(5), 1011–1038 (1989)

Smith, O.J., Westerberg, A.W.: Mixed-integer programming for pressure swing adsorption cycle scheduling. Chem. Engng. Science 45(9), 2833–2842 (1990)

Vsevolodov, N.: Biomolecular Electronics, an Introduction via Photosensitive Proteins. Birkhaeuser, Boston (1998)

Yang, R.T.: Adsorbents: Fundamentals and Applications. Wiley & Sons, Hoboken, NJ (2003)

Zhang, T.H., Zhang, C.P., Fu, G.H., Li, Y.D., Gu, L.Q., Zhang, G.Y., Song, Q.W., Parsons, B., Birge, R.R.: All-optical logic gates using bacteriorhodopsin films. Optical Engng. 39(2), 527–534 (2000)

Chapter 6
Self-Evolvability for Biosystems

Abstract. The straightforwardness with which biosystems solve complex problems suggests adopting the strategies developed in nature to face evergrowing complexity for other systems.

Hypercubes for genetic code, hypercycles as a principle of self-organization, and NK-models of evolution describing genotype fitness landscape are presented.

The hierarchy of structure, function, dynamics, within spatial and temporal brain scales is characterized by the K-set models.

The correlation with differential models, entropy criteria and bio-inspired computing methods as, autonomic, and organic computing is presented.

6.1 Hypercube for Genetic Code

Self-evolution is a first criterion for biology. The structural capacity of biosystems to self-evolve is based on the genetic code.

Several hypothetic scenarios have been advanced to explain the genetic code structure and its origin (Koonin and Novozhilov 2009).

The expanding genetic code scenario from single-base nucleotides to doublets and then to triplets, that is to codons, offers interesting suggestions for self-evolvability studies and applications for higher complexity systems.

The main theories on origin and evolution of the code are the stereo-chemical theory, the co-evolution theory and the adaptation theory.

The stereo-chemical theories suppose that the specificity of a codon for a particular amino acid is based on a direct interaction of amino acid and nucleotides. Amino acids might have been binding directly to the codons when the code was established and such binding imposed the code. The co-evolution theory explains the non-randomness of the code by the fact that the code system is an imprint of the prebiotic pathways of amino-acid formation. According to this theory the genetic code evolution reflects the relationship among amino acids and their biosynthesis. An early code used fewer codons and amino acids and then expanded to include new amino acids arising from biosynthesis coded for by new codons, with the resulting code assigning similar codons to amino acids that are related by biosynthesis.

Adaptive codes theories attempted to explain the observed patterns in genetic code and its evolution by postulating optimality of the code. Adaptation theories state that selection pressure resulted in the emergence of a code optimized for some measure, such as for minimization of the physicochemical effects of single mutational or translational errors.

These theories suggest that the genetic codes are information communication system that should reflect the physico-chemical properties of the amino acids. The different theories are not mutually exclusive and probably the code was shaped by a compromise of several scenarios (Ardell and Sella 2002).

The genotype of cells is laid down in a linear sequence of four nucleotides: A-adenine, C-cytosine, U-uracil and G-guanine. The genetic information is transcribed in messenger RNA, mRNA, used as instructions for protein translation. Translation requires a mapping of the four nucleotides in 20 amino acids. Triplets of the four different RNA bases are read sequentially from mRNA. DNA is transcribed to mRNA that makes use of an RNA adaptor, transfer RNA, tRNA to interpret nucleotides in amino acids. The four bases, C, G, U and A, might form 64 different simple triplets patterns, the so-called codons. The 20 amino acids and the start and stop signals are coded redundantly by these 64 codons.

The symmetry elements in genetic code supported the use of algebraic frames to characterize code origin and function. It has been suggested that the overall layout of the code can be accurately described in the algebra of group theory or of fields (Findley et al. 1982, Jimenez-Sanchez 1995. Jimenez-Montano et al. 1996, Jimenez-Montano 1999). These symmetries may be of help in explaining regularities and periodicities as observed in proteins sequences. They have been correlated to the possible evolution scenarios of the genetic code.

The relevant group to describe the symmetries of the bases{C, G, U, A} should be a group of order 4. There are only two possibilities for the group structure, the cyclic group

C (4) and the group associated to the Galois field, GF (4). This is the Klein 4-group.

Several codes can be associated to the genetic code according to the order of importance for bases and of their positions in codons.

For triplets or codons the ranking: position 2>position 1>position 3 in establishing the amino acid is acknowledged.

One of the proposed nucleotide hierarchical ordering is: C>G>U>A. This hierarchy was established starting from the observation that C, in position 2 in codon, is anytime able to be source of a single amino acid. G is able to determine the amino acids in majority of cases, U only in some cases and A never. In other words, C base passes any time a single message, while U and A are credited with at least double message. G passes stronger messages than U or A, concerning the coded amino acid.

It is possible to associate to any base in codon a two-digit vector: [hydrogen bonds, chemical nature]. The first digit refers to hydrogen bonds and the second to the chemical nature. We will use first digit "1" for high number of hydrogen bonds that is for G and C and second digit "1" for chemical nature pyrimidines that is for C and U.

We will use first digit “0” for low number of hydrogen bonds that is for A and U and second digit “0” for chemical nature purines that is for A and G.

In this way we may describe the basis by the two-digit vectors:

C= [11]. G= [10], U= [01], A= [00].

This corresponds to the hierarchy: C>G>U>A, and to the real numbers 3, 2, 1, and 0 associated to C, G, U, and A since 0= [0 0], 1= [01], 2= [10], 3= [11].

The four nucleotides C, G, U, A may be represented as a 2-cube, as shown in Fig. 6.1.

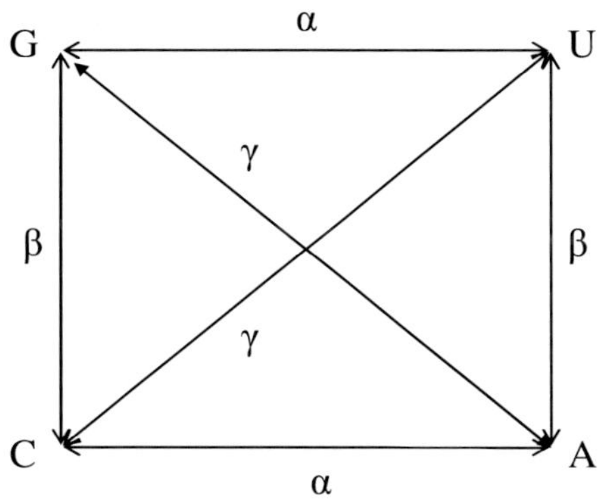

Fig. 6.1 The 2-cube of nucleotides

Here α denotes the transversions between non-complementary bases, β denotes the transversions between complementary bases and γ denotes the transitions (Bertman and Jungck 1979).

Of course, restricting the nucleotide characterization to only two properties that is hydrogen bonds and chemical nature, is a drastic simplification.

The polytope of genetic code could be derived as a solution of the wave equation, WE that is eq. 3.4 and eq. 3.5. The elements of the WE, as Y, T, Z, Q, are considered as polytopes.

Particular solutions of the general WE are given by the kinetic eq. (3.6).

Considering Y (0) = 0 it results in the solution.

$$Y= Q \otimes T \tag{6.1}$$

The detailed form of this solution depends on the structures associated to Q, T and to the product between them.

The successive values of T are [00], [01], [10] and [11] associated to 0, 1, 2 and 3 and it has the structure of group.

Let us suppose that Q has the same structure as T.

If Q and T are structured as Klein 4-groups and the product is the Cartesian product of these, we obtain Y as a hypercube (Bertman and Jungck, 1979, Jimenez-Montano et. al 1996).

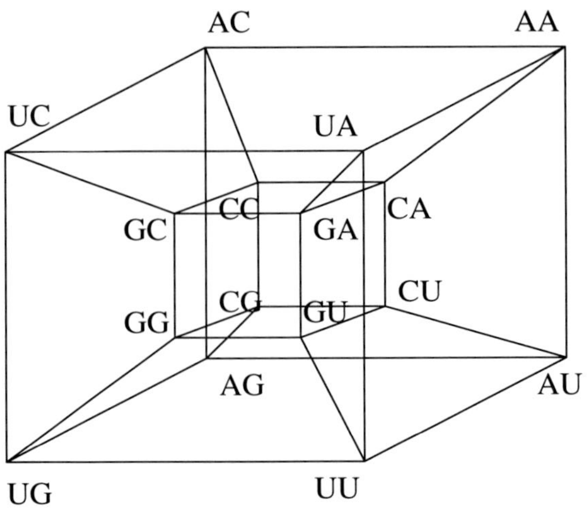

Fig. 6.2 Group graph polytope

According to Bertman and Jungck the genetic code doublets can be divided into two octets of completely degenerate and ambiguous coding dinucleotides. These two octets have the algebraic property of lying on continuously connected planes on the group graph, of the Cartesian product of two Klein 4-groups of nucleotide exchange operators. The product group can also be broken into four cosets, one of which has completely degenerate coding elements, and another that has completely ambiguous coding elements. The two octets of coding doublets have the further algebraic property that the product of their internal exchange operators naturally divides into two exactly equivalent sets. These properties of the genetic code are relevant to unraveling error detecting and error correcting, that is proofreading, aspects of the genetic code and may be helpful in understanding the context-sensitive grammar of genetic language.

Fig. 6.2 shows the group graph polytope associated to the 16 doublets of the genetic code (Bertman and Jungck 1979).

Fig. 6.3 shows the 4-cube as the polytope associated to the 16 doublets of the genetic code (Jimenez-Montano et. al 1996).

Fig. 6.4 shows the proposed here logical polytope for doublets. It is based on the logical hypercube (Moretti 2009). We associated the value false, F=0 and true, T=1 for edges of the logical hypercube. It resulted in 16 connectives of 4 digits where the doublets have been identified.

Observe that the proposed here polytope as shown in Fig. 6.4 differs from that discussed in the literature (Bertman and Jungck 1979, Jimenez-Montano et al. 1996).

Bertman and Jungck considered as basic transformations, α and β instead of β and γ, as we did. Since α changes two nucleotides, we do not consider it as basic.

The proposed hypercube differs also from that studied by Jimenez-Montano.

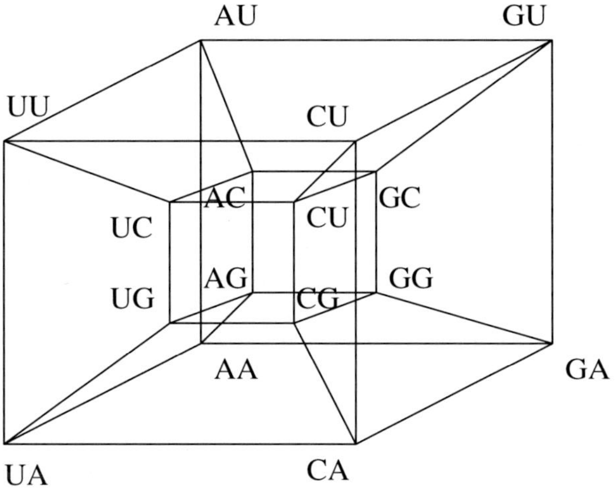

Fig. 6.3 The 4-cube for doublets

In the proposed here model the transition from external to internal cube is of the type β while for Jimenez-Montano model it is of the type γ.

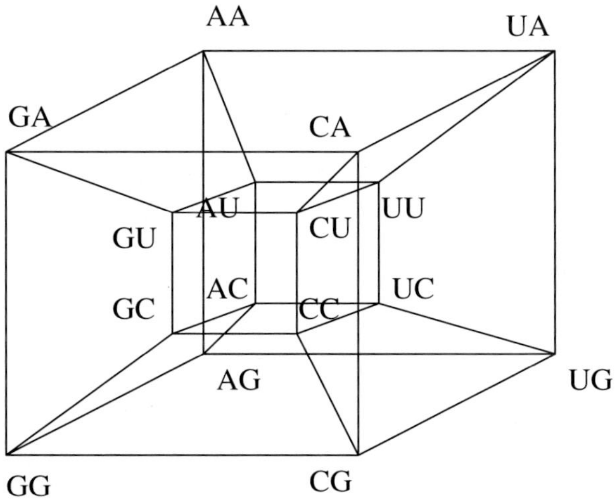

Fig. 6.4 Logical polytope for doublets

Fig. 6.5 based on Fig. 6.4 outlines the two types of doublets. The fat-line polygon shows the doublets with strong coding nucleotides, C and G. The dot-line polygon shows the doublets with weak coding nucleotides, U and A.

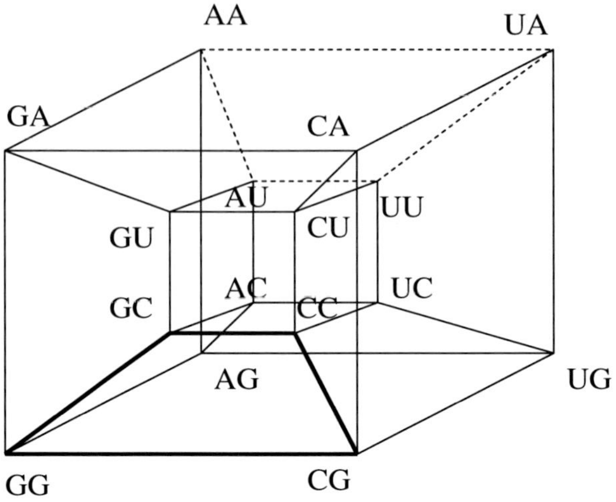

Fig. 6.5 Logical polytope for doublets and coding

Table 6.1 shows the doublet codings.

The relation with the group of strong coding and weak coding doublets as shown in Fig. 6.5 is taken into account.

Table 6.1 shows that the strong doublets are able to code for only one amino acid.

The same is valid for the mixed doublets of the internal cube.

The weak doublets and the mixed doublets of the external cube code are shown for more than one amino acid.

The analysis suggests that there are two complementary roots for genetic code evolution.

One consists in introducing two letters C and G which promoted higher information capability and higher physico-chemical stability.

On the other side there are A and U coding for seven amino acids and a Stop.

Coupling both ways allows genetic diversity, versatility and evolution.

Table 6.1 Doublet coding and amino acids

		Third letter		
	Doublet	A or G		U or C
Strong	CC		Pro	
	CG		Arg	
	GC		Ala	
	GG		Gly	
Mixed intern	CU		Leu	
	GU		Val	
	UC		Ser	
	AC		Thr	
Mixed extern	CA	Gln		His
	GA	Glu		Asp
	UG	Trp(Stop)		Cys
	AG	Arg		Ser
Weak	UU	Leu		Phe
	UA	Stop		Tyr
	AU	Met		Lys
	AA	Lys		Asn

6.2 Entropy Criteria for Genetic Code

Starting from the fact that C= [11]. G= [10], U= [01], A= [00], the strong doublet rectangle in Fig. 6.5 has a digit representation: CC= [1111], CG= [1110], GC= [1011] and GG= [1010].

The set of four doublets corresponds to a 4x4 matrix and based on this we can use entropic analysis.

To this matrix one may associate a similarity matrix. The entropy associated to this matrix is H (Strong) = 6.475.

By adding new compounds to this reference mixture of four doublets, the entropy H varies.

There is only a small change of entropy, ΔH if the vector of the test compound is similar to the reference set and this supplementary compound is thought to have similar properties.

If a database shares similar bit patterns with reference set molecules, adding a similar compound will induce a change targeting the minimum entropy production.

By contrast, inclusion of a doublet compound having dissimilar vector leads to a higher entropy production, targeting the maximum entropy production.

In this way database compounds may be screened to identify a compound that causes low or high changes of the reference set informational entropy and detects other promising drug according to the established goal.

The tested directions are AC= [0011], UC= [0111], GU = [1001], CU= [1101], GA= [1000], CA= [1100], AG= [0010] and UG= [0110].

Denote by Strong the matrix of strong doublets.

$$\text{Strong} = \begin{vmatrix} 1 & 1 & 1 & 1 \\ 1 & 1 & 1 & 0 \\ 1 & 0 & 1 & 1 \\ 1 & 0 & 1 & 0 \end{vmatrix} \tag{6.2}$$

The informational entropy calculations are shown in Table 6.2.

Table 6.2 shows the informational entropy for strong doublets.

Denote by Weak the matrix of weak doublets.

$$\text{Weak} = \begin{vmatrix} 0 & 0 & 0 & 0 \\ 0 & 0 & 0 & 1 \\ 0 & 1 & 0 & 0 \\ 0 & 1 & 0 & 1 \end{vmatrix} \tag{6.3}$$

Table 6.2 Informational entropy for strong doublets

New Step	Strong	Strong [0011]	Strong [0111]	Strong [1101]	Strong [1101]	Strong [1000]	Strong [1100]	Strong [0010]	Strong [0110]
Matrix	1111 1110 1011 1010	1111 1110 1011 1010 0011	1111 1110 1011 1010 0111	1111 1110 1011 1010 1001	1111 1110 1011 1010 1101	1111 1110 1011 1010 1000	1111 1110 1011 1010 1100	1111 1110 1011 1010 0010	1111 1110 1011 1010 0110
H	6.475	11.958	11.958	10.894	10.894	10.894	10.894	11.958	11.958
ΔH	0	5.483	5.483	4.419	4.419	4.419	4.419	5.483	5.483
Δ^2H	-	0	-1.064	0	0	0	1.064	0	-

Table 6.3 shows the informational calculus for transition toward the same doublets from the direction of the Weak doublets.

Table 6.3 Informational entropy for weak doublets

New step	Weak	Weak [0011]	Weak [0111]	Weak [1001]	Weak [1101]	Weak [1000]	Weak [1100]	Weak [0010]	Weak [0110]
Matrix	0000 0001 0100 0101	0000 0001 0100 0101 0011	0000 0001 0100 0101 0111	0000 0001 0100 0101 1001	0000 0001 0100 0101 1101	0000 0001 0100 0101 1000	0000 0001 0100 0101 1100	0000 0001 0100 0101 0010	0000 0001 0100 0101 0110
H	6.475	10.894	10.894	11.958	11.958	11.958	11.958	10.894	10.894
ΔH	0	4.419	4.419	5.483	5.483	5.483	5.483	4.419	4.419
$\Delta^2 H$	-	0	1.064	0	0	0	-1.064	0	-

A kind of periodicity for $\Delta^2 H$ can be observed. This suggests looking for periodicities in code evolution.

The genetic code symmetry may be correlated to aminoacid similarity.

It is expected that amino acids corresponding to similar codons will have similar physical–chemical properties.

The problem is to define similarity or distance between two amino acids.

Different definitions of similarity and distance from Section 2.5 may be of use.

Some similarities have been outlined in Table 6.4.

Amino acids like His/Gln or Asp/Glu are very close.

Table 6.4 Table of similar amino acids

Amino-acids	Codons	First Digit Different
Ser/Arg	AGC/AGG	6
Cys/Trp	UGC/UGG	6
Phe/Leu	UUC/UUG	6
Ile/Met	AUC/AUG	6
His / Gln	CAC/CAG	6
Asp/Glu	GAC/GAG	6
Tyr/Stop	UAC/UAG	6
Asn/Lys	AAC/AAG	6
Pro/Ser	CCC/UCC	3
Cys/Trp	UGC/UGG	6
Leu/Val	CUC/GUC	4
Ala/Thr	GCC/ACC	3

Comparable pairs of amino acids have been evaluated by previous models (Frappat et al. 2000).

Similarity between codons has practical applications since the associated closely related codons might be replaced without significant effects on the function of the protein (Benyo et. al. 2004). The most frequently occurring replacements are described by different amino-acids substitution group tables (Taylor 1986).

According to Table 6.4 the distance between His and Gln is lower than that between Leu and Val and this in turn lower than that between Ala and Thr or between Pro and Ser.

The parametrization presented here challenges in part that offered by models where pairs of amino acids as Leu and Val, or Pro and Ser, or Ala and Thr are considered as strongly similar (Frappat et al. 2000). The strong similarity is unfounded in these cases since we are faced with hydrophobic and hydrophilic amino acids in the same pair.

6.3 Hypercycles

A notable approach in the study of genetic code evolution is the Eigen's model of hypercycles systems of mutually autocatalytic components. It considers the question of under what conditions, the system can self-organize to a dynamic stability (Eigen 1971, Eigen and Schuster1979). The approach was based on the view that the self-organization including the development of hypercycles is a process that can occur in a homogeneous system by intrinsic necessity.

Eigen and Schuster proposed the model of hypercycles as a hypothetical stage of macromolecular evolution, which could follow the quasispecies.

The hypercycle is a self-reproducing macromolecular system, in which RNAs and enzymes cooperate in the following manner: there are RNA matrices (I_i); ith RNA codes i-th enzyme E_i ($i = 1,2,...,n$); the enzymes cyclically increase RNA's replication rates, namely, E_1 increases replication rate of I_2, E_2 increases replication rate of I_3, ..., E_n increases replication rate of I_1. In addition, the mentioned macromolecules cooperate to provide primitive translation abilities, so the information, coded in RNA-sequences, is translated into enzymes, analogously to the usual translation processes in biological objects. The cyclic organization of the hypercycle ensures its structure stability. For effective competition, the different hypercycles should be placed in separate compartments.

The replication enzymes ensure the more accurate RNAs' replication as compared with quasispecies, providing opportunities for further macromolecular structure improvements. Eigen and Schuster consider hypercycles as predecessors of protocells, the primitive unicellular biological organisms.

Developing the hypercycle model, Eigen and Schuster discussed the difficult problem of how could the real very complex translation mechanism and unique genetic code be created during macromolecular self-organizing process. Plausible evolution steps were outlined and a corresponding well-defined mathematical model was developed.

Eigen and Schuster considered that the primitive genetic code may use units of less than three bases. During its early evolution, the code would have increased both the number of codons and the coded amino acids and the present code would reflect the pattern of this historical expansion (Wilhelm and Nikolajewa 2004).

In the view of Kuhn and Waser, understanding the origin of living systems is an engineering problem: to find a sequence of physicochemical stages, beginning with prebiotically reasonable conditions and leading to self-organization of matter and to systems equipped with a life-like genetic apparatus (Kuhn and Waser 1994, Kuhn and Kuhn 2003).

The genome generates different dynamical systems that promotes their stability and survive and in that way serves as seeds of a generally self-evolvable system.

The genome may be interpreted as a possible solution of the wave equation, WE, model.

It is an apparently timeless model since the time T is defined on a finite group and has a cyclic character. As discussed by H. Kuhn this temporal cycling is crucial for genetic code emergence and evolution. Dynamical model with usual real time, characterizing the kinetic equations, completes the self-evolvable system description.

The interplay between the wave equation, WE, that is, eq. 3.4 and eq. 3.5, in the so-called sequence space and the real-valued equations of thermodynamics and chemical kinetics corresponds to the specificity of living systems.

Fig. 6.6 illustrates the cyclic schemas associated to hypercycles.

In this model the conditions K are associated to RNA while the states S to enzymes.

The hypercycle is a self-reproducing macromolecular in which RNAs and enzymes cooperate. There are RNA matrices (I_i), the ith RNA codes the ith enzyme E_i. The enzymes cyclically increase RNAs' replication rates, namely, E_0 increases replication rate of I_1, E_1 increases replication rate of I_2, E_2 increases replication rate of I_3, and E_3 increases replication rate of I_0.

The mentioned macromolecules cooperate to provide primitive translation abilities, so the information, coded in RNA-sequences, is translated into enzymes analogous to the usual translation processes in biosystems.

The cyclic organization of the hypercycle ensures its structure stability. For effective competition, the different hypercycles should be placed in separate compartments.

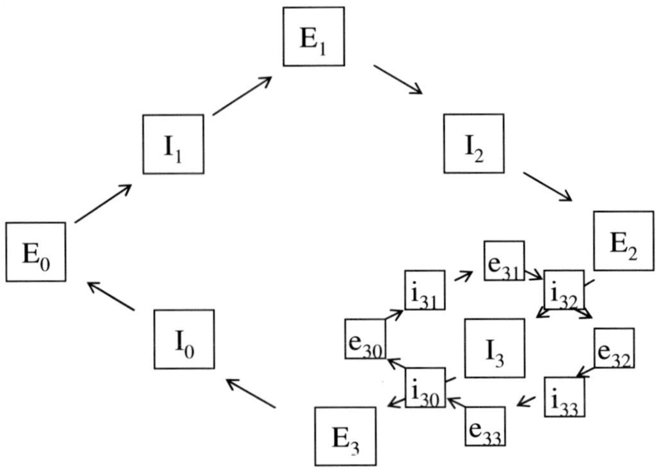

Fig. 6.6 Schemas for hypercycles

Fig. 6.6 shows that some RNAs may induce the reproduction of other metabolites in cyclic manner. Supposing that I_3 is in this situation, e_{30} increases replication rate of i_{31} , e_{31} increases replication rate of i_{32}, e_{32} increases replication rate of i_{33}, and e_{33} increases replication rate of i_{30}. The number of RNAs in each cycle may vary.

The quasispecies mathematical model was put forward by Eigen and Peter Schuster based on the initial work done by Eigen (Eigen 1971).

The correlation between the quasispecies model and the wave equation WE is meaningful.

Suppose that there are n different nucleic sequences $x_1,\ldots,x_n$, with replication rates $a_1,\ldots,a_n$. These quantities represent the selective values of the individual mutants.

In the absence of mutations the variant with the highest replication rate will grow fastest and reach fixation. The result of selection in a world without errors is a homogeneous population consisting of the fastest replication variant.

But replication is not error-free. Thus it is necessary to define the probability q_{ij} that erroneous replication of template x_j results in the production of the sequence x_i.

The quantities q_{ij} form the n x n mutation matrix.

If we consider binary sequences and point mutations, we obtain:

$$q_{ij} = p^{H_{ij}}(1-p)^{(L-H_{ij})} \tag{6.4}$$

Here p is the mutation rate per bit, L is the length of the bitstring and H_{ij} is the Hamming distance between the strains i and j, that is the number of bits in which the two strains differ. Error-free replication is given by:

$$q_{ii} = (1-p)^{L} \tag{6.5}$$

The quasispecies equation is given by:

$$\frac{dx_i}{dt} = \sum_{j=1}^{n} a_j q_{ij} x_j - \phi x_i \tag{6.6}$$

The variants x_j replicate at rate a_j, and generate mutants x_i, with probabilities q_{ji}.

The death term Φx_i is chosen to keep the total population size constant.

This is the case if Φ denotes the average fitness of the population given by:

$$\phi = \sum_{i=1}^{n} a_i x_i \tag{6.7}$$

The relation between the quasispecies equation and the wave equation, WE is significant.

The wave model, WE, characterizes the genetic bio-chemical reactor in a discrete space. It includes the convection or drift term $V \otimes \frac{\partial Y}{\partial Z}$ and the kinetic term $Q \otimes Y$.

The convection term corresponds to mutation in quasispecies equation while the kinetic term corresponds to selection and death term.

It could happen that the convection contribution is more significant than that of selection for evolution.

Observe that just one wave equation, WE, aims to replace the entire system of differential equations for quasispecies (Eigen and Schuster 1979).

WE is adequate for highly non-linear processes modeling. The time T is a more natural expression for time to record qualitative developments than the usual linear time. The cyclic and diversified characters of environment, as described by Kuhn, are accounted for by T and Z introduced here. Different values of T correspond to the developmental or pattern recognition stages.

The quasispecies model is a description of the process of the Darwinian evolution of certain self-replicating entities within the framework of physical chemistry. In other words, a quasispecies is a large group or cloud of related genotypes that exist in an environment of high mutation rate, where a large fraction of offspring are expected to contain one or more mutations relative to the parent. This is in contrast to a species, which from an evolutionary perspective, is a more-or-less stable single genotype, most of the offspring of which will be genetically accurate copies.

It is mainly useful in providing a qualitative understanding of the evolutionary processes of self-replicating macromolecules such as RNA or DNA or simple asexual organisms such as bacteria or viruses, and is helpful in explaining something of the early stages of the origin of life. Quantitative predictions based on this model are difficult because the parameters that serve as its input are hard to obtain from actual biological systems.

Fig. 6.7 illustrates the utility of polytopes to describe evolutionary dynamics with phenotype (Schuster 2002).

A cross-polytope, or orthoplex, is a regular, convex polytope that exists in any number of dimensions. The cross-polytope is the convex hull of its vertices. Its facets are simplexes of the previous dimension, while the cross-polytope's vertex figure is another cross-polytope from the previous dimension.

Fig 6.7 shows the formation of a new variant, I_n, through mutation of a genotype present in the population. The polynucleotide sequence is processed through to yield the corresponding phenotype S_n. The phenotype, in turn, is evaluated by a mapping which returns fitness relevant properties in quantitative terms. These values appear in the dynamical system as parameters of the new species. Eventually the new variant is fully integrated into the replication mutation ensemble.

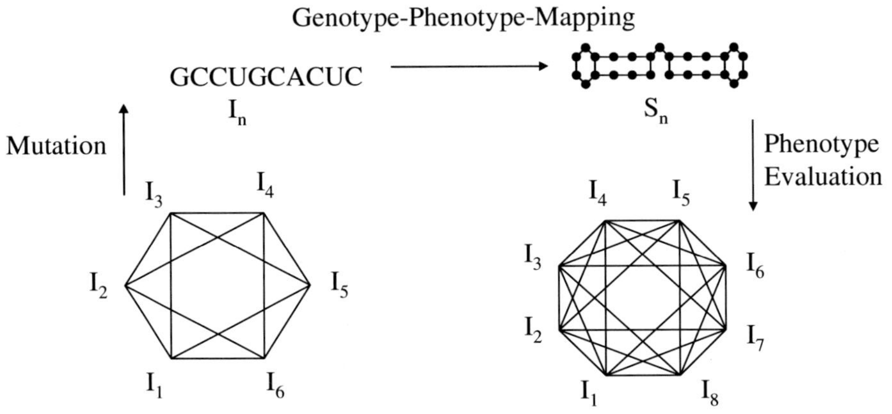

Fig. 6.7 Evolutionary dynamics with phenotype

Observe that the two hypercycles, shown in Fig. 6.7, are cross-polytopes.

Fig. 6.7 describes is a transition from tricross (3-orthoplex) to tetracross (4-orthoplex).

6.4 Sequence Space

One way to study the diverse nucleotide sequences in the genes of viruses is to map them into a multidimensional matrix called a sequence space (Eigen 1993).

In this space, each point represents a unique sequence, and the degree of separation between points reflects their degree of dissimilarity. The space can be most easily drawn for short sequences consisting of binary digits. For a sequence with just one position, there are only two possible sequences, and they can be drawn as the end points of a line. For a sequence with two positions, there are four permutations, which form the corners of a square. The variations on a three-digit sequence become the corners of a cube, and the variations on a four-digit sequence are the vertices of a four-dimensional hypercube.

Fig. 6.8 illustrates two steps for sequence construction, that corresponding to cube and hypercube. Each higher-dimensional space is built iteratively by drawing the previous diagram twice and connecting the corresponding points.

The sequence spaces for viral genomes are far more complex than these simple figures because they involve thousands of positions that can each be occupied by one of the four different nucleotides.

The construction of a high-dimensional sequence space was illustrated by Eigen's studies (Eigen et al. 1988). Each additional sequence position adds another dimension, doubling the diagram for the shorter sequence.

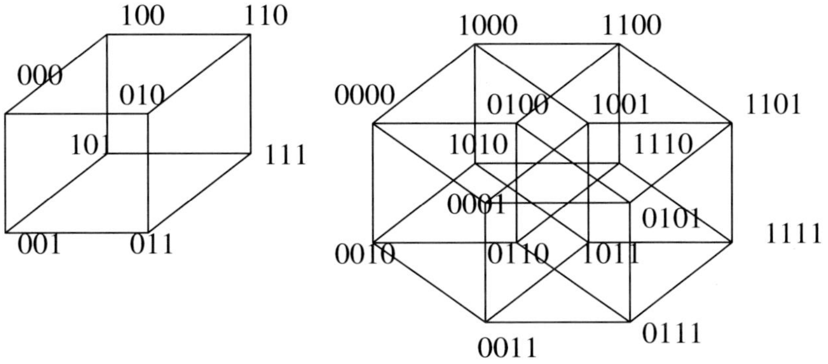

Fig. 6.8 Sequence space

The progression is from a cube to a hypercube. A four, or twenty, letter code can be accommodated either through allowing four, or twenty, values for each dimension or through additional dimensions.

This may be written as a hypercube.

It should be emphasized that the wave equation WE may generate these hypercubes.

Fig. 6.9 shows a projection of the polytope of sequence space.

The interpretation of viruses as quasispecies (Eigen 2000, Sole et al. 2006) allows understanding viruses as n-categories and suggests new strategies to confront the viruses.

The population dynamics for quasispecies may be described by fractal polytopes (Pearse 2007).

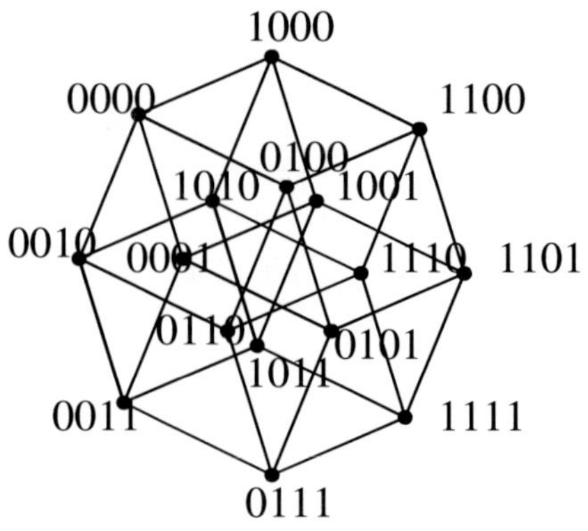

Fig. 6.9 Polytope for sequence space

Fractal polytopes are fractals formed by repeated replacing a polytope with smaller polytopes at its vertices.

This family of fractals contains many fractal constructions such as the Menger sponge.

A self-similar system is a family Φ_j of contraction similitude.

The Menger sponge is constructed via the maps:

$$\phi_j(x) = \frac{x}{3} + p_j \tag{6.8}$$

where p_j (a_j, b_j, c_j) for a_j, b_j, $c_j \in \{0, 1/3, 2/3\}$, except for the six cases when exactly two coordinates are 1/3, and the single case when all three coordinates are 1/3.

Population dynamics of a quasispecies depend on the error rate of its replication process. Fig. 6.10 contains highly simplified representations of the sequence spaces that might contain a quasispecies. If the replication process of a quasispecies were perfectly accurate, all the quasispecies offspring would occupy the same position in sequence space. If replication were highly imperfect, mutant quasispecies would soon occupy every position in sequence space, and the quasispecies population would lose its integrity. This corresponds to Fig. 6.10a.

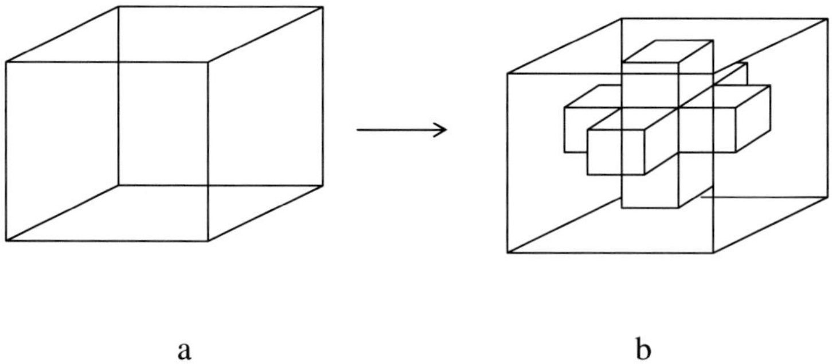

Fig. 6.10 Self-similar tilling for quasispecies

At some intermediate error rate, however, the quasispecies population would become a coherent, self-sustaining entity that resembles a cloud centered on the original consensus sequence. This corresponds to Fig. 6.10b. The cloud shape represents the quasispecies.

6.5 Self-Fabrication

The self-fabrication refers to a process that happens spontaneously without assistance from a fabricator.

The concept of autonomous self-fabrication was developed in the theory of self-reproducing automata based on universal constructors (Von Neumann 1966).

Von Neumann showed that self-fabrication of a machine, that is, the autonomous turnover of self on the basis of a supplied blueprint and self-reproduction, making a copy of self, including the blueprint, that is the design, is in principle possible.

The so-called kinematical self-reproducing machine consists of a general-purpose fabricator P +Φ (X), which is an automaton consisting of two parts: a constructor P that fabricates a machine X from spare parts according to Φ (X), the blueprint for X.

When supplied with its own blueprint Φ (P), the constructor makes itself.

To give the entire fabricator P + Φ (X) the ability to make a copy of itself, Von Neumann added a blueprint copier Q and a controller R, so that the fabricator becomes (P + Q + R) + Φ (X), which can make not only X but also a copy of Φ (X).

When supplied with its own blueprint, Φ (P + Q + R), it can make a copy (P + Q + R) + Φ (P + Q + R) of itself and of its blueprint, thereby ensuring self-fabrication of the full system (Hofmeyr 2007).

Von Neumann discusses logical arguments for the necessity of symbols as distinct from dynamics in self-replication. The motivation for his argument was to understand complex systems and to design computers, but his basic conclusions apply to existing cellular replication and to self-fabrication systems.

As an attempt to describe the physical support for Von Neumann automata, Pattee evaluated the properties of memory, codes, symbolic control, and material construction that would promote efficient evolutionary search and natural selection (Pattee 2005).

In even the simplest existing cells the steps from the symbolic base sequence in DNA to a functioning enzyme are too complex to have originated without simpler intermediate stages. However, to control construction or synthesis, even the simplest one-dimensional discrete-state memory storage that exists by virtue degenerate energy states, must somehow control the rates of specific dynamical interactions. This means that the linear degeneracy must be broken. This must be done by new interactions between the linear storage elements. In existing cells this is a complex process that requires several steps. First, the DNA sequence is transcribed to messenger RNA, mRNA, by template copying. Next the coding enzymes and transfer RNAs, tRNA, translate the base triplet code to the corresponding amino acids that are then joined in sequence by the messenger

RNA and ribosome machinery. Finally the one-dimensional sequence folds into a functioning enzyme. In this process there are cases of descriptions and constructions by both template inspection and coded descriptive translations.

The discovery of enzymatic RNA made it possible to imagine a simpler translation process in which RNA can function both as a constructing enzyme and as a symbolic description of an enzyme. The description is considered as a passive structure that can be copied by template inspection, while the construction is a dynamic catalytic process that joins molecules by strong, covalent bonds. The main point is that this double function is only possible by virtue of the two configurations of RNA, the passive one-dimensional sequence memory and the folded three-dimensional active ribozyme.

It has been pointed out that ribosomes are the only known examples of Von Neumann constructors (Hofmeyr 2007). The ribosome corresponds to the Self. They fit the description entirely: on its own a ribosome can do nothing, but in conjunction with the information embedded in the mRNA molecule that has been transcribed from DNA it can string amino acids together in the specified sequence. This is done with the help of auxiliary enzymes, cofactors and an energy source. However, the genetic blueprint for a ribosome is made up of a set of individual blueprints for the numerous protein and ribonucleic acid components that make up a ribosome. There is no contiguous genetic blueprint for a complete ribosome. Therefore, a ribosome never directly makes a ribosome, only the protein bits from which it is made up. The ribosomal RNAs are made by ribosomally-synthesized enzymes.

Consequently the problem of whether a Von Neumann constructor can fabricate itself directly does not arise in the cell. However, we still need to explain how the ribosomal components assemble into a fully functional entity. The fabrication of all ribosomes entails two processes: the construction of the parts that is, the polypeptide chains and ribosomal RNA, and their subsequent assembly into a fully functional entity.

There is another process wedged in between, namely that of the folding of newly synthesized polypeptide chains into a functional, three-dimensional conformation.

Folding transformations are the most significant semiotic process in all living systems. Folding is fundamental because it is the process that transforms the passive symbolic informational sequences into dynamic rate-controlling constraints. Physically to describe folding in any structure requires two types of bonds, strong bonds that preserve the passive topological structure of what is folded, and weaker bonds that acting together hold the active folded structure in place. As long as the strong-bond topological sequence structure is energy degenerate it can serve as an informational constraint or a passive memory. Folding removes this degeneracy by allowing new weak-bond interactions between the elements resulting in an active enzyme. A physical description of protein folding is an energy minimization process or a relaxation of many weak-bond interactions under the constraints of the strong bonds holding the linear sequence together (Wolynes et al. 1995).

Fig. 6.11 shows a possible association between the elements of a self-fabrication in a cell and the general PSM frame. Ribosomes appear in Fig. 6.11 as biophysico-chemical example of what means the Self. Obviously other identifications may exist.

The substrate S refers to nutrients. We could identify K1 as tRNA, K2 as DNA and K3 as mRNA. A deconstruction way should be considered too for artificial systems.

This means that after the integration way, S→K1→K2→K3, we need to look at the differentiation way, K3′→K2′→K1′→S′. The swinging between both ways will allow self-fabrication.

By disassembly, system decomposes itself into subsystems or parts. In this way an association which is not necessary any more may disassemble or may be disconnected.

For self-fabrication, the disassembly process allowing the formulation of new systems is as important as the self-assembly itself.

Assembly and disassembly are dual concepts, and need algebra and dual algebra for characterization.

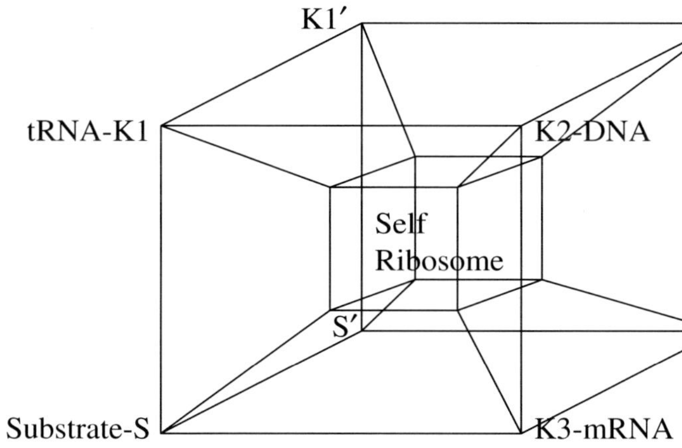

Fig. 6.11 Polytope for self-fabrication

The main objective of the polytope shown in Fig. 6.11 is to understand and to make use of similar architectures as suggestion for artificial self-fabricating systems.

6.6 NK-Model for Biosystems

The NK-model was introduced by Kauffman as a problem independent model for constructing fitness landscapes that can gradually be tuned from smooth to rugged (Kauffman 1993).

The main parameters of the model are N, the number of genes in the genotype, that is the length of the strings that form the points in the landscape, and K the number of other genes that epistatically influence a particular gene. The fitness contribution of each gene is determined by the gene itself plus K other genes.

Several properties of the model are independent of A, the number of possible values each gene can have, so the simplest case A=2 may be used.

The fitness of a bit string b of length N is defined as follows. Suppose that to every bit b_i, i=1,…,N, in a bit string b is assigned a fitness f_i, of its own.

The fitness f_i of a bit b_i does not only depend on the value, 0 or 1, of the bit itself, but also on the value of K other bits b_j in the same bit string. These dependencies are called epistatic interactions.

So the fitness contribution of one bit depends on the K+1 bits, itself and K others, giving rise to a total of 2^{K+1} possibilities called neighborhood configurations. Each of these neighborhood configurations is assigned a random fitness value. Therefore, the fitness contribution f_i of a bit b_i is specified by a list of 2^{K+1} values. This assignment of fitness values is repeated every bit b_i, i=1,…, N in the bit string b. There are N lookup tables, one for each bit, and each with 2^{K+1} entries.

The fitness of the entire bit string is defined as the average fitness contribution of all the bits:

$$F = \frac{1}{N}\sum_{i=1}^{N} f_i \tag{6.9}$$

Let us consider the case N=3 and K=2.

Fig. 6.12 shows the eight genotypes arranged in a Boolean cube.

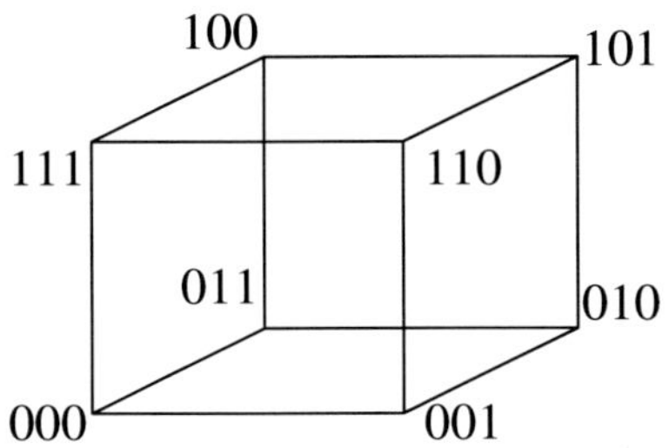

Fig. 6.12 NK-model**:** Boolean cube frame

The arrangement of the eight genotypes allows a trajectory in the natural increasing order of numbers from 0=000 to 7=111 associated to genotypes.

The lookup table for N=3 and K=2 contains 24 values.

The NK-models show that fitter genotypes move at greater heights than less fit genotypes. Another example to consider is the genotype with only four genes, each having two alleles, 1 and 0 that is, a Boolean representation of the state of each gene, resulting in 16 possible genotypes, each a unique combination of the different states of the four genes (Fig. 6.13). Each vertex differs by only one mutation from the neighboring vertices, representing the step of a single mutation, thereby showing that each mutation as such is independent of the state of the other genes.

An adaptive walk begins at any vertex, moves to vertices that have higher fitness values and ends at a local optimum, the vertex that has a higher fitness value than all its one-mutant neighbors. For one case study, three local optima exist where adaptive walks may end. In random landscapes, looking for the global peak by searching uphill is useless; it is the same as searching the entire space of possibilities (Kauffman 1995).

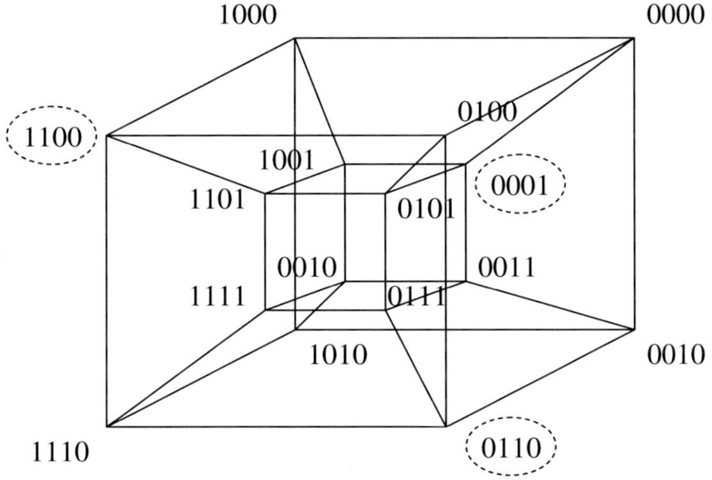

Fig. 6.13 NK-model: 16 possible peptides

Sixteen possible peptides 4 aminos long are arranged as vertices on a four-dimensional Boolean hypercube. Each peptide connects to its four one-mutant neighbors, accessible by changing a single amino acid from 1 to 0 or from 0 to 1. The hypercube represents this four-dimensional peptide space. Each peptide has been assigned, at random, the rank-order fitness, ranging from the worst, 1, to the best, 16. Directions of such moves between adjacent positions are shown by arrows from the less fit to the more fit. For the example considered by Kauffman, local optimums correspond to 1100, 0001 and 0110 as highlighted in Fig. 6.13 (Kauffman 1993).

In reality, the fitness landscapes that underlie the mutation steps of gradualism are correlated, and local peaks do often have similar heights. Through the existence of particular evolutionary phenomena, developmental pathways, regulatory genes and epigenetics, no gene exists on its own; all genes correlate to other genes; this is often referred to as epistatic coupling or epistatic interactions.

Rugged landscapes are those landscapes in which the fitness of one gene depends on that one part and upon K other parts among the N present in the landscape.

Building on this, the NK-model offers further insight into the mechanisms of evolution and selection (Kauffman 1993). Again, consider an organism with N gene loci, each with two alleles, 1 and 0. Let K stand for the average number of other loci, which epistatically affect the fitness contribution of each locus.

The fitness contribution of the allele at i locus depends on itself, whether it is 1 or 0, and on the other alleles, 1 or 0, at K other loci, hence upon K+1 alleles. The number of combinations of these alleles is just 2^{K+1}. Kauffman selects at random from each of the 2^{K+1} combinations a different fitness contribution from a uniform distribution between 0.0 and 1.0. The fitness of one entire genotype can be expressed as the average of all of the loci. Generally, epistatic interactions create a more deformed landscape.

Despite the importance of fitness landscapes for evolutionary processes, the landscapes may vary from smooth, single-peaked to rugged, multi-peaked. During evolution, species search these landscapes using mutation, recombination and selection, a process for which the NK-model provides insight into particular phenomena accompanying the adaptive walk.

These fitness landscapes have already been used in the context of networks.

Worth mentioning is the work which shows that searches are most likely to be more effective for combining technologies rather than those for new technologies; this finding indicates firms collaborating by combining technologies might have more success than those that search solely for new technologies.

The main difficulty in utilizing NK-models is the need to introduce a large number of adjustable parameters for fitness evaluation. Unless the parameters possess a precise physical meaning, the modeling becomes an exercise in landscape curve generating while some important qualitative feature of the phenomenon could be lost within the numerical simulations.

For this reason a new type of differential model for fitness is presented.

It follows the structure developed for wave equation, WE, model.

Denote by F (Z) the fitness, by Z the space genotypes and by Q the kinetic rate.

The model characterizes the fitness variation along Z.

It is supposed that F varies at a constant rate Q along Z.

The model is:

$$\frac{\partial F}{\partial Z} \oplus Q \geq 0 \tag{6.10}$$

$$F(0) = G \tag{6.11}$$

$F=f_0f_1f_2\ldots f_j$, $Z=z_0z_1z_2\ldots z_j$, $Q=q_0q_1q_2\ldots q_j$, $G=g_0g_1g_2\ldots g_j$ where fj, zj, qj and gj are digits. The initial fitness function is G.

A solution in C (2) is presented here for illustration purposes.

F, Z, Q, G are defined in C (2). This means that we refer here to cyclic processes.

In C (2), "0" denotes the null element. The null value "0" may be interpreted as that corresponding to a non-activated scale while the unit value "1" corresponds to an activated scale of the system.

The real product and the sum were translated to C (2) operations. The solution of the equality in (3.42) similar to Euler solution for differential equations will be:

$$F(Z) = F(0) \oplus Q \otimes Z \tag{6.12}$$

Suppose that F (0) =G=111. In this case the solution (3.44) of the balance equation

F (Z, Q) is shown in Table 6.5.

Table 6.5 shows the fitness values.

Table 6.5 Fitness F (Z)

Q\Z	000	001	010	011	100	101	110	111
000	1	1	1	1	1	1	1	1
001	1	0	1	0	1	0	1	0
010	1	1	0	0	1	1	0	0
011	1	0	0	1	1	0	0	1
100	1	1	1	1	0	0	0	0
101	1	0	1	0	0	1	0	1
110	1	1	0	0	0	0	1	1
111	1	0	0	1	0	1	1	0

According to Table 6.5 the rate Q takes the minimum non-null value, Q=001. This corresponds to the situation in which only one scale is involved in the process and to minimum production of roughness.

The rate Q=000 corresponds to no roughness, Q=001 to one scale roughness and Q=110 to high roughness based on two scales.

The maximum production of roughness corresponds to Q=111 and is based on the three possible scales.

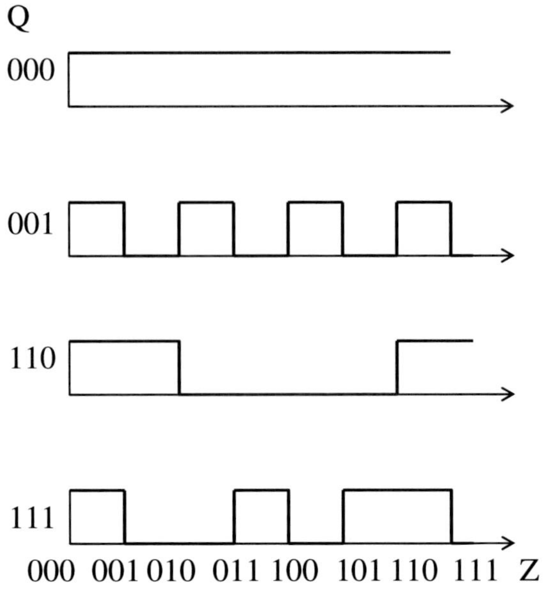

Fig. 6.14 Fitness F (Z)

The resulting fitness is shown in Fig. 6.14.

6.7 K-Set Model

The hierarchy of structure, function, dynamics, within spatial and temporal brain scales is described by the K-set models. K-set models are mesoscopic, that is, intermediate scale models, introduced by Freeman in the 1970s. They represent an intermediate level of hierarchy between microscopic neurons and macroscopic brain structures (Freeman 1975).

K-set are multi-scale models, able to describe increasing complexity of structure and dynamical behavior.

The basic building block is the K0 set which describes the dynamics of a cortical micro-column with about 10.000 neurons. K-set models allow topological specifications of the hierarchy of connectivity in neuron populations in the 6-layer cortex. A KI set contains K0 sets from a given layer with specific properties. KII includes KI units from different populations, that is, excitatory and inhibitory ones. KIII has several KII sets modeling various cortical areas. KIV covers cortical areas across the hemisphere. KV is the highest level of hierarchy describing neocortex. The dynamics of K-set has the following hierarchy: K0 has zero fixed point attractor; KI has non-zero fixed point attractor; KII has limit cycle oscillations; KIII exhibits chaos; and KIV shows intermittent spatio-temporal chaos. The function of KIII sets can be sensory processing and classification using a single channel; KIII may correspond to visual sensory system, olfactory system, hippocampus, midline forebrain, and so on. KIV performs multisensory fusion and decision making. KV has components of higher cognition and conscious functions.

KV set that deals with the highest level functions in mammalian cognition, which have yet to be formally assessed navigational and perceptual features have been modeled using KIV sets.

The KV set was proposed as a model of the unique properties of neocortex, which maintains multiple unstable periodic orbits that appear as overlapping phase cones. It is postulated that the robust background activity of the neocortex manifests the continuous engagement of the organism with its environment. Proofs using non-linear differential equations and possibly new forms of mathematics, as for instance the n-category theory, are required to integrate the K set into an organic whole.

KV is proposed to model scale-free dynamics in mammalian cognition and is still underdeveloped. In order to categorize KV we may exploit the notion of categorical colimit, as the object that acts as the glue of patterns of neural connectivity.

The neocortex and KV appear as biophysico-chemical living proof of the Self.

Table 6.6 shows the basic elements of the K-set hierarchy. It refers to the categorification of the K-set hierarchy.

The K-set model may be interpreted in the general PSM frame.

Comparing the two models we may associate S to K0, K1 to KI and KII, K2 to KIII, K3 to KIV and the Self to KV. Obviously other identifications may be considered.

Three types of sensory signals are considered in KIV: exteroceptors, interoceptors, including proprioception, and orientation signals; for instance, gravity, visual flow, magnetic fields. Each of these sensory signals provides stimuli toward the brain, namely the sensory cortices, midline forebrain, MF, unit, and the hippocampal formation, HF, respectively. The model is not intended to mimic all the biological details; rather it is used to incorporate the main elements required for operation of brains at the KIV level of functionality.

Another KIII component of the integrated KIV system, the Midline Forebrain, MF, formation, receives the interoceptor signals through the basal ganglia, and processes them in the hypothalamus and the septum. MF provides the value system of the KIV, using information on the internal goals and conditions in the animal. It provides the information stream to the amygdala, which combines this with information coming from the cortex and the hippocampus to make a decision about the next step/action to be taken.

The motor part of the model limbic system is driven by the simulated amygdala. The direction of motion that it determines is based on the combined information from the three sensory systems, which collectively form the architecture of the global KIV. From EEG studies we infer that a cooperative state emerges from the collective interaction among the CA1, PC, Septum, and Amygdala, by which various behavioral patterns are formed and executed. The model is designed to provide the platform with which to study by simulation this behavior formation and action selection mechanism.

Table 6.6 Categorification for K-set hierarchy

Type	K0	KI	KII	KIII	KIV	KV
Polytope	S	K1	K1	K2	K3	Self
-	n=0	n=1	n=2	n=3	n=4	n=5
Category	0-cat	1-cat	1-cat	2-cat	3-cat	4-cat
Structure	Single Unit	Populations excitatory inhibitory units	Interacting populations	Several interacting KII and KI sets	Interacting KIII sets	Integration previous K set
Dynamics	Non-linear I/O function	Fixed point converges to zero or non-zero value	Fixed point converges to zero or non-zero value	Aperiodic, chaotic oscillation	Spatio-temporal dynamics with global phase transitions	Multiple unstable periodic orbits that overlap phase
In brain	All higher level K sets of K0 units	PG, DG, BG, BS	OB, AON, PC, CA1, CA3, CA2, HT, BG, BS, AMY	Cortex, HC, M F	Hemisphere cooperation cortical, HF and MF by AMY	Neocortex

* Notations: PG-periglomerular; OB-olfactory bulb; AON-anterior olfactory nucleus; PC-prepyriform cortex; HF-hippocampal formation; DG-dentate gyrus; CA1, CA2, CA3-cornu ammonis sections of the HC-hippocampus; MF-midline forebrain; BG-basal ganglia; HT-hypothalamus; DB-diagonal band; SP-septum, AMY-amygdala, BS-brain stem.

The elements of K set may be associated to the polytope from Fig. 6.15.

The notations are: S-Perception, K1-Interoceptors, K2-Orientation Beacon, and K3-Motor Skills.

The central stage, the Self, ensures the cooperation and redistribution of the four stages on another face of the polytope, with another starting stage.

Any stage embeds the previous ones. After one cycle an augmented reality may support a new cycle of development.

As shown in Fig. 6.15 the swinging between closure and disclosure way should be considered too.

This means that after the integrative way S→K1→K2→K3 we need to look at the differentiation way K3′→K2′→K1′→S′.

Making use of the developments of the direct way may result in a kind of symmetry-breaking for the new result. The swinging from direct to reverse developmental stages mediated by the Self, is critical for creativity in complex problem solving since the boundaries where creativity grows consist of synchronized, integrative and differentiation tendencies (Engstrom and Kelso 2008, Kelso and Tognoli 2009). The need for both ways finds a support in the studies of metastable coordination dynamics of the brain.

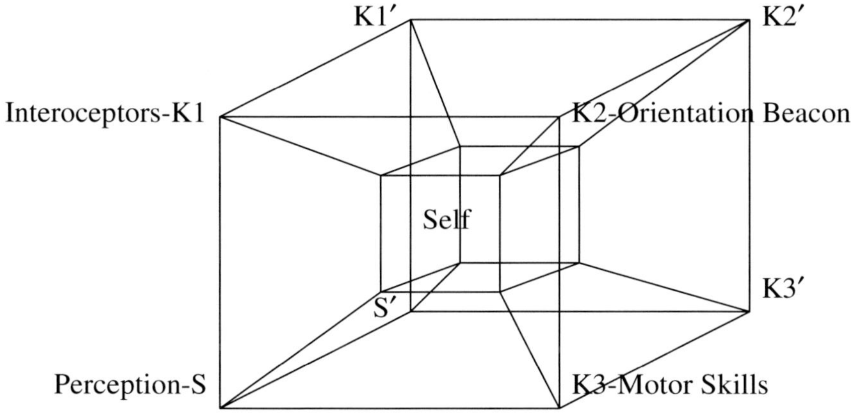

Fig. 6.15 Polytope for K-set model

Some ideas about brain organization have emerged that may provide support for polytopic representation of the cognitive processes. One step in this direction is the complementarist epistemology and ontology (Ji 1995).

Ji draws on the biology of the human brain, namely, the dual nature of its hemispheric specializations. The left and right hemispheres have relatively distinct psychological functions, and the ultimate reality, as perceived and communicated by the human brain, is a complementary union of opposites (Ji 1995).

On a much finer-grained scale, Stephen Grossberg (2000) has drawn attention to the dual, nature of brain processes and explained how the brain is functionally organized to achieve self-adaptive behavior in a changing world.

Grossberg (2000) presents one alternative to the computer metaphor suggesting that brains are organized into independent modules. Evidence is reviewed that brains are organized into parallel processing streams with dual properties. Hierarchical interactions within each stream and parallel interactions between streams create coherent behavioral representations that overcome the complementary deficiencies of each stream and support unitary conscious experiences. This perspective suggests how brain design reflects the organization of the physical world with which brains interact.

For example, the visual system is divided by virtue of its sensitivity to different aspects of the world, form and motion information being carried by ventral and dorsal cortical pathways. The working memory order is dual to working memory, rate the color processing is dual to luminance processing, and so forth. The brain is organized this way in order to process dual types of information in the environment. A goal of research was to study more directly how complementary aspects of the physical world are translated into dual brain designs for coping with the world (Grossberg 2000).

Freeman outlined metastability role in neocortex activity (Freeman and Holmes (2005).

A comparison of the Freeman and Kelso approaches to non-linear brain dynamics shows that:

- Both Freeman's and Kelso's approaches appeal to Haken's synergetics and to so-called circular or reciprocal causality
- Both approaches appeal to non-linear coupling among neural oscillators as the basis for varying degrees of global integration
- Both theories make use of basin attractor dynamics to interpret experimental data
- Both approaches invoke symmetry-breaking coordination dynamics from the loss of attractors of the relative phase dynamics

Metastable coordination dynamics it is not about states but about the accurate mixing of integration and differentiation tendencies.

The role of the swinging or wave behavior of the system should be emphasized (Freeman 2007). The oscillatory regime is an important goal for creative systems that can autonomously find solutions to highly complex and ill-defined construction problems. This corresponds to the wave type of behavior.

6.8 Autonomic Computing

Over the past years technical systems as vehicles, airplanes, telecommunication networks, manufacturing systems, became more and more complex. This is the result of the embedding of hardware and software into these systems.

The term selfware has been coined to refer to the set of self-properties that are emerging in the autonomic or organic computing.

The evergrowing complexity of today's IT systems has led to unsustainable increases in their management and operation costs. Hardware and software architects and developers aim to alleviate this problem by building self-evolvable, or autonomic, systems, that is, systems that self-configure, self-optimize, self-protect and self-heal based on a set of high-level, user-specified objectives.

With respect to the future evolution new advanced management principles have to be developed. A feasible principle is an autonomic behavior of the system which is addressed by two significant research directions, namely autonomic and organic computing. Autonomy with its reference to a self, or autos, refers to an independence from external influences of different sort (Sterritt and Hinchey 2005).

Biologically inspired autonomic and organic computing systems are essentially concerned with creating self-directed and self-managing systems based on suggestions from nature and the human body, such as autonomic nervous system.

Autonomic computing (Kephart and Chess 2003) is a computing initiative that draws analogies from the autonomic nervous system where all reactions occur without explicit override by the human brain-so to say autonomous. By embedding this behavior into technical systems, the complexity can be left to the systems themselves. One refers to this autonomy as self-properties. This means self-configuration, that is, configuration and reconfiguration according to policies, self-optimization, that is, permanent improvement of performance and efficiency, self-healing, that is, reactive and proactive detection, diagnostics and reparation of localized problems and self-protection, that is, defense of the system as a whole. Furthermore, autonomic computing systems are expected to be self-aware, context sensitive, anticipative and adaptive.

The implementation of autonomic computing concept offered preliminary suggestions for the study of self-integrative closure problem in self-evolvable control systems.

At the heart of an autonomic system is a control system, which is a combination of components that act together to maintain actual system attribute close to desired specifications.

An autonomic system embodies more closed loops. The standard autonomic computing system can be modeled in terms of two main control loops, local and global, with sensors for self-monitoring, effectors for self-adjustment, knowledge and planer/adapter for exploiting policies based on self-evolvability and environment awareness.

It should be noted that similar architectures are of interest for both autonomic and organic computing (Trumler et al. 2004, Bauer and Kasinger 2006).

As software systems become increasingly complex and difficult to manage, the autonomic computing was developed as a way of handling this. Software should actively manage itself instead of passively being managed by a human administrator. Most self-properties can be achieved under the responsibility of a single autonomous entity, a manager, which controls a hierarchy of other autonomous entities. The autonomic manager consists of a central loop, which handles all upcoming events within the system. The autonomic manager follows the MAPE loop, which stands for monitoring, analysis, planning and execution, supported by a knowledge base.

The logical structure of an autonomic element is similar to that of self-evolvable BDI agents. For autonomic computing, the agent structure is replaced by the so-called MAPE loop whose elements are M-Monitor, A-Analyze, P-Plans, and E-Execute.

Autonomic computing systems shown in Fig. 6.16 are composed of four levels that may be identified as K0 or S-Managed Resources, K1-Touchpoints, K2-Touchpoints Autonomic Managers, K3-Orchestrated Autonomic Managers. To this a central Manual Manager is to be considered. The closed loop in which K3 is replaced by an automatic device was also studied.

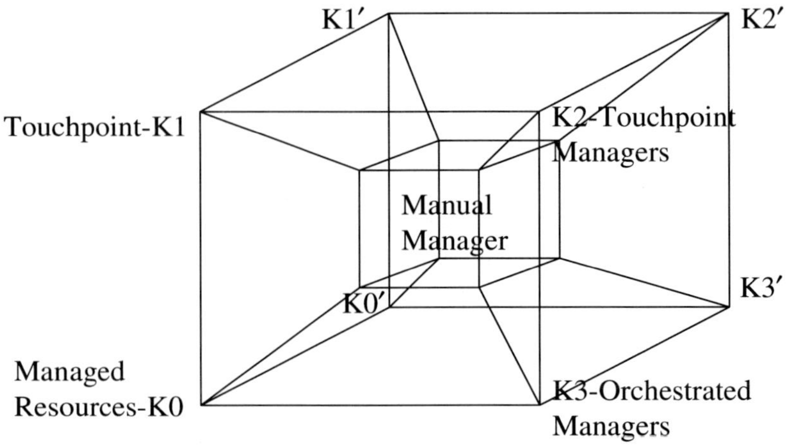

Fig. 6.16 Polytope for autonomic computing

Fig. 6.16 shows the main elements of the automatic computing architecture.

Calinescu and Kwaitowska proposed a computer-aided development of self-evolvable systems (Calinescu and Kwaitowska (2009). The corresponding polytopic development is shown in Fig. 6.17.

The notations for Fig. 6.17 are: K0-Sensors, Monitors, K1-Analyze, K2-Plan, and K3-Execute Effectors.

The entire system is managed by self-knowledge system model.

Fig. 6.16 and Fig. 6.17 suggest that after the integration, or direct epistemology way S→K1→K2→K3 we need to look at the differentiation, or reverse epistemology way K3′→K2′→K1′→S′.

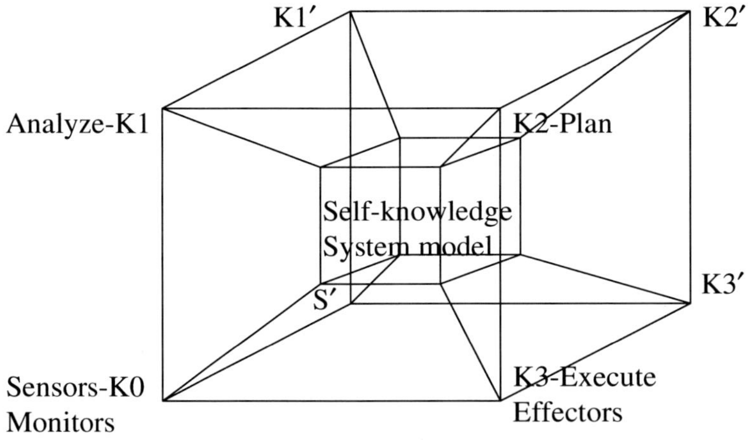

Fig. 6.17 Polytope for self-evolvable system

The reverse way appears as a balance to the centralized approach. Decentralized autonomic computing is beneficial since interacting and fairly autonomous individuals may complete the manager activity. The new information is created by the concomitant tendencies, direct and reverse.

6.9 Organic Computing

Organic computing is a research field emerging around the conviction that problems of organization in complex systems in computer science, telecommunications, neurobiology molecular biology, and ecology can be presented in a unified way, by means of which progress in understanding aspects of organization in either field can be fruitful in the others.

Problems of organization become pressing as artifacts increase in complexity regarding both hardware and software. It is becoming inevitable to shift much of the burden of organization into the machines themselves. This brings up the problem of keeping their self-organization controllable. This requires interfaces for user interaction on a high level, which hides the rise in inner complexity from the users. An organic computing system is a technical system, which adapts dynamically to the current conditions of its environment.

From the computer science point of view, the apparent easiness with which living systems solve computationally difficult organizational problems makes it inevitable to adopt strategies observed in nature for creating information processing machinery.

Organic computing investigates the design and implementation of self-organizing systems that are self-configuring, self-optimizing, self-healing, self-protecting, self-describing, context aware, and anticipatory. Thus, organic computing includes the autonomic computing targets. Organic computing emphasizes on biological and organic-inspired systems and on the aspects of self-organization and emergence. Meeting the grand challenge of organic computing requires scientific and technological advances in a wide variety of fields.

Organic computing system draw analogies from living systems and try to use perceptions about the functionality of living systems for the development and management of artificial and technical systems. In addition to the self-properties of autonomic computing systems, they are defined as being self-organizing. This is a critical step to ensure autonomy.

Organic computing is a project which combines software engineering with neuroscience and molecular biology (Würtz 2008). Within this framework, the controller observer architecture was developed to keep emergent behavior within predefined limits (Schloer and Muller-Schloer 2005).

It allows the system to make free decisions within so-called adaptive islands, limited by preset objectives and constraints.

The basic structure consists of an execution unit which receives an input and generates an output. Above the execution unit, there is an observer/controller unit. The observer receives input from the environment as well as from the execution unit. The controller compares the situation reported by the observer to the goals set by the user and reacts by reconfiguring the execution unit.

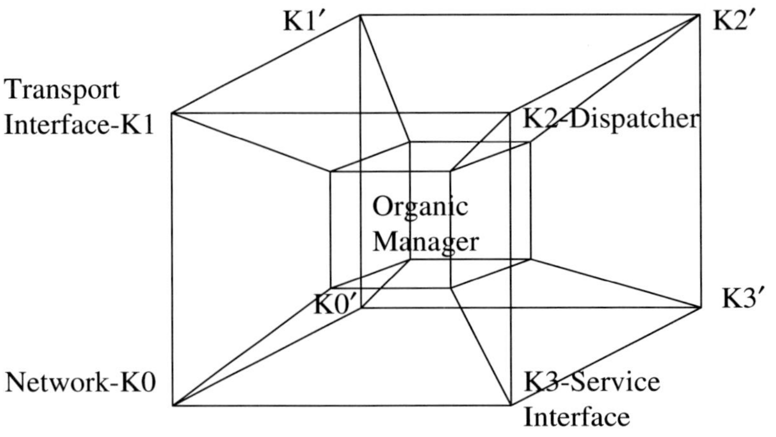

Fig. 6.18 Polytope for organic computing

For the organic computing middleware architecture, the levels may be identified as: K0 or S as the Network, K1-Transport Interface, K2-Event Dispatcher, K3-Service Interface and Proxy (Trumler et al. 2004). To these an Organic Manager is joined.

In the middleware architecture the organic manager is linked to the previous levels and ensures a meta-representation of them. This is critical for self-organization.

Fig. 6.18 shows the automatic computing architecture.

The notations are: K0-Network, K1-Transport Interface, K2-Dispatcher, and K3-Service Interface

The two-way approach is suggested by Fig. 6.18.

This means that after the direct way of integration K0→K1→K2→K3 we are looking for the reverse way of differentiation K3′→K2′→K1′→K0′.

A two-level organic manager should be able to correlate the two ways.

The reverse way makes use of the developments of the direct way. This may explain the different results for reverse way. The need of both direct and reverse ways is critical.

Neither way has enough explanatory and predicting value without the other.

Only both ways, in duality, ensure self-evolvability.

That is because the boundary where creative research grows and new information is created is that of synchronized tendencies. Tendencies to integrate should coexist with tendencies to differentiate and it is the timing and balance of both that matters for self-evolvability.

Successes of autonomic and organic computing have been reported in the fields of drug and new materials discovery, data communications, computer animation, control and command, exploration systems for space, undersea and harsh environments and there exists much promise for future progress.

References

Ardell, D.H., Sella, G.: No accident: genetic codes freeze in error-correcting patterns of the standard genetic code. Philos. Trans. R. Soc. London B. Biol. Sci. 357(1427), 1625–1642 (2002)

Bauer, B., Kasinger, H.: AOSE and organic computing-how can they benefits from each other. Position paper, AOSE III. Springer (2006)

Benyo, B., Biro, J.C., Benyo, Z.: Codes in the codons: Construction of a codon/amino acid periodic table and a study of the nature of specific nucleic acid-protein interactions. In: Proceedings of the 26th Annual International Conference of the IEEE EMBS, San Francisco, pp. 2860–2863 (2004)

Bertman, M.O., Jungck, J.R.: Group graph of the genetic code. The J. of Heredity 70, 379–384 (1979)

Calinescu, R., Kwiatkowska, M.: CADS*: Computer-Aided Development of Self-* Systems. In: Chechik, M., Wirsing, M. (eds.) FASE 2009. LNCS, vol. 5503, pp. 421–424. Springer, Heidelberg (2009)

Engstrom, D., Kelso, J.: Coordination dynamics of the complementary nature. Gestalt Theory 30(2), 121–134 (2008)

Eigen, M.: Selforganization of Matter and the Evolution of Biological Macromolecules. Naturwissenschaften 58(10), 465–523 (1971)

Eigen, M.: Viral Quasispecies. Scientific Am. 269, 42–49 (1993)

Eigen, M.: Viruses: Evolution, Propagation and Defense. Nutrition Reviews 58(2), 5–16 (2000)

Eigen, M., Schuster, P.: The hypercycle a principle of natural self-organization. Springer, Berlin (1979)

Eigen, M., McCaskill, J., Schuster, P.: Molecular quasispecies. J. Phys. Chem. 92, 6881–6891 (1988)

Findley, G.L., Findley, A.M., McGlynn, S.P.: Symmetry characteristics of the genetic code. Proc. Natl. Acad. Sci. USA 79, 7061–7065 (1982)

Frappat, L., Sciarrino, A., Sorba, P.: Theoretical Prevision of Physical-Chemical Properties of Amino Acids from Genetic Code (2000), `http://arxiv.org/list/physics/0007034`

Freeman, W.J.: Mass Action in the Nervous System. Academic Press, New York (1975)

Freeman, W.J.: Neurodynamics. An Exploration of Mesoscopic Brain Dynamics. Spinger, London UK (2000)

Freeman, W.J.: Proposed cortical "shutter" mechanism in cinematographic perception. In: Perlovsky, L., Kozma, R. (eds.) Neurodynamics of Cognition and Consciousness, pp. 11–38. Springer, Heidelberg (2007)

Grossberg, S.: The complementary brain: A unifying view of brain specialization and modularity. Trends in Cognitive Sciences 4, 233–246 (2000)

Freeman, W.J., Holmes, M.D.: Metastability, instability, and state transition in neocortex. Neural Networks 18(5-6), 497–504 (2005)

Hofmeyr, J.-H.: The biochemical factory that autonomously fabricates itself: a systems biological view of the living cell. In: Boogard, F., Bruggeman, F., Hofmeyr, J.-H., Westerhoff, H. (eds.) Systems Biology: Philosophical Foundations. Elsevier (2007)

Ji, S.: Complementarism: A biology-based philosophical framework to integrate western science and eastern tao. In: Proceeding of the 16th International Congress of Psychoterapy, pp. 518–548 (1995)

Jimenez-Montano, M.A.: Protein evolution drives the evolution of the genetic code and vice versa. Biosystems 54, 47–64 (1999)

Jimenez-Montano, M.A., de la Mora-Basanez, R., Poschel, T.: The hypercube structure of the genetic code explains conservative and non-conservative amino acid substitutions in vivo and in vitro. Biosystems 39, 117–125 (1996)

Jimenez-Sanchez, A.: On the Origin and Evolution of the Genetic Code. J. Mol. Evol. 41, 712–716 (1995)

Kauffman, S.A.: The origins of order: Self-organization and selection in evolution. Oxford University Press, New York (1993)

Kauffman, S.: At home in the universe: The search for laws of self-organization and complexity. Oxford University Press, New York (1995)

Kelso, J.A.S., Tognoli, E.: Toward a Complementary Neuroscience: Metastable Coordination Dynamics of the Brain. In: Murphy, N., Ellis, G.F.R., O'Connor, T. (eds.) Downward Causation and the Neurobiology of Free Will. UCS, vol. 3, pp. 103–124. Springer, Heidelberg (2009)

Kephart, J.O., Chess, D.M.: The vision of autonomic computing. IEEE Computer 36(1), 41–50 (2003)

Koonin, E.V., Novozhilov, A.S.: Origin and evolution of the genetic code: The universal enigma. IUBMB Life 61(2), 99–111 (2009)

Kuhn, H., Kuhn, C.: Diversified world: drive of life's origin? Angew. Chem. International 42, 262–266 (2003)

Kuhn, H., Waser, J.: Hypothesis on the origin of genetic code. FEBS Letters 352, 259–264 (1994)

Moretti, A.: The Geometry of Logical Opposition. PhD. Thesis. University of Neuchâtel, Switzerland (2009)

Patee, H.: The Physics and Metaphysics of Biosemiotics. Journal of Biosemiotics 1, 281–301 (2005)

Pearse, E.P.J.: Canonical self-similar tilings by iterated function systems. Indiana Univ. Math. J. 56, 3151–3170 (2007)

Schöler, T., Müller-Schloer, C.: An Observer/Controller Architecture for Adaptive Reconfigurable Stacks. In: Beigl, M., Lukowicz, P. (eds.) ARCS 2005. LNCS, vol. 3432, pp. 139–153. Springer, Heidelberg (2005)

Schuster, P.: A testable genotype-phenotype map: Modeling evolution of RNA molecules. In: Läassig, M., Valleriani, A. (eds.) Biological Evolution and Statistical Physics, pp. 56–83. Springer, Berlin (2002)

Sole, R.V., Sardanyes, J., Díez, J., Mas, A.: Information catastrophe in RNA viruses through replication thresholds. J. Theor. Biol. 240, 353–359 (2006)

Sterritt, R., Hinchey, M.: Autonomic Computing – Panacea or Poppycock? In: The 12th IEEE International Conference and Workshops on the Engineering of Computer-Based Systems, pp. 535–539 (2005)

Taylor, W.R.: The classification of amino acid conservation. J. Theor. Biol. 119, 205–218 (1986)

Trumler, W., Bagci, F., Petzold, J., Ungerer, T.: Towards an organic middleware for the smart dooplate project. GI Jahrestagung (2), 626–630 (2004)

Von Neumann, J.: Theory of Self-Reproducing Automata. University of Illinois Press, Urbana (1966)

Wilhelm, T., Nikolajewa, S.L.: A new classification schema of the genetic code. J. Mol. Evol. 59, 598–605 (2004)

Wolynes, P., Onuchic, J.N., Thirumalai, D.: Navigating the folding routes. Science 267, 1619–1620 (1995)

Würtz, R.P. (ed.): Organic Computing: Series: Understanding Complex Systems. Springer (2008)

Chapter 7
Self-Evolvability for Cognitive Systems

Abstract. The post-formal and closure aspects for cognitive developmental stages, geometry of logic, and relational complexity theories are presented.

Conceptual and computational frameworks are presented as polytopic cognitive architectures.

Physarum computing capabilities are evaluated.

7.1 Developmental Stages

Cognitive structures are patterns of physical or mental actions that underlie specific acts of intelligence and correspond to the stages of development (Piaget 1970, 1971).

According to Piaget, there are four primary cognitive development stages: sensory-motor, preoperational, concrete operational and formal.

Fig. 7.1 shows the developmental stages hierarchy.

It was observed that restriction of cognitive capability to the formal stage may correspond to systems stagnation and unavoidable failure (Yang and Bringsjord 2005).

This refers to automata that have a code or protocol that recommend some actions for situations requiring a completely different code.

Growing complexity imposes to look for creativity and self-evolvability for automata.

Piaget's epistemology made room for cognition beyond the fourth stage. Piaget initiated the study of post-formal stages, beyond the fourth, in which agents are able to operate over logical systems. This refers to meta-processing of logics and formal theories expressed in those logics. It was considered that elaboration of axiomatic schemas may be considered as surpassing the formal stage and are to formal schemas what the latter are to concrete operations (Piaget 1973).

The post-formal stages appeared as possible candidates for the so-called 5th cognitive development stage (Bringsjord et. al 2010). They are comparable to the formal framework in which post-formal reasoning involves the Self.

Fig. 7.2 shows a polytopic presentation of the cognitive developmental stages.

The initial four stages of Piaget, associated to S, K1, K2, and K3, have been completed in Fig. 7.2 by the self-evolvability stage. This allows describing systems able to self-evolve by internal structures modification.

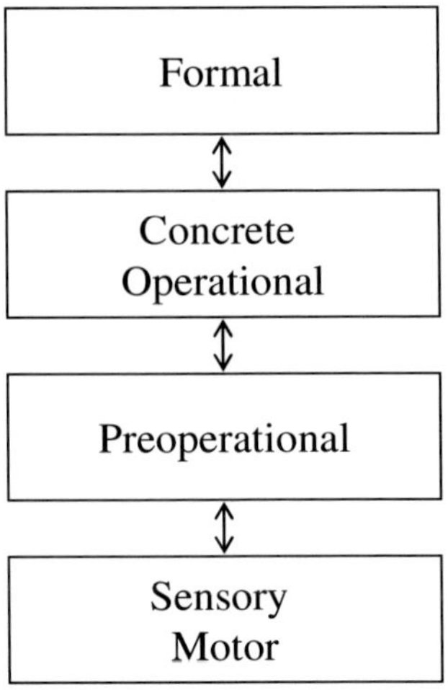

Fig. 7.1 Developmental stages hierarchy

There are four stages on the front face of the polytope. The notations are: S-Sensory Motor, K1-Preoperational, K2-Concrete Operational, and K3-Formal.

The development is considered clockwise.

Piaget considered that the sensorimotor stage differed from the latter stages in that the former was devoid of symbolic representation.

The central stage the Self may ensure the cooperation and redistribution of the four stages on another face of the polytope, with another starting stage.

Any stage embeds the previous ones. After one cycle an augmented reality may support a new cycle of development. The post-formal stage appears as a cognitive exemplar of the Self.

As shown in Fig. 7.2 two ways should be considered for development.

This means that after the integrative way S→K1→K2→K3 we need to look at the differentiation way K3′→K2′→K1′→S′.

Using the developments of the direct way may produce symmetry-breaking results for the reverse way. The swinging from direct to reverse developmental stages mediated by the Self may be a source of creativity in complex problem solving or science development.

That is because the boundaries where creative research stand out and new information is created consist of coexisting tendencies. Integration and differentiation coexists and the metastable coordination dynamics emerges as the delicate blend of integration and differentiation tendencies.

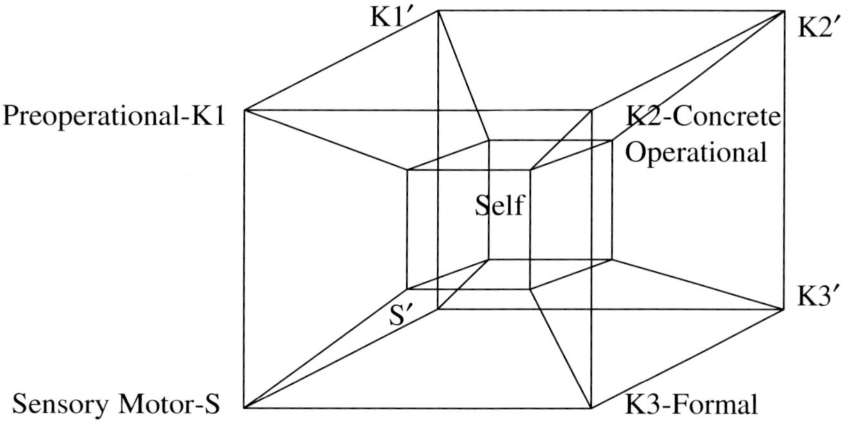

Fig. 7.2 Polytope for development stages

7.2 Logical Polytope

The Boolean logic operations may be illustrated by a polytope whose vertices represent the 16 traditional binary connectives that is, logical operations on two variables, of basic logic (Moretti 2009).

Table 7.1 shows the binary propositional connectives.

Table 7.1 Binary propositional connectives

⊤	∨	←	**P**	→	**q**	↔	∧	**NAND**	**XOR**	**¬q**	**N→**	**¬p**	**N←**	**NOR**	⊥
T	T	T	T	T	T	T	T	F	F	F	F	F	F	F	F
T	T	T	T	F	F	F	F	T	T	T	T	F	F	F	F
T	T	F	F	T	T	F	F	T	T	F	F	T	T	F	F
T	F	T	F	T	F	T	F	T	F	T	F	T	F	T	F

The binary-connective labels in Table 7.1 correspond to the digital labels shown in Fig. 7.3. Thus the binary-connective labels and the digital labels provide different ways of looking at the same abstract structure, which can itself be interpreted either as a Hasse diagram of a Boolean lattice or as a polytope. Table 7.1 shows the 16 connectives. We associate T to the digit "1" and F to the digit "0".

Fig. 7.4 shows a different presentation of the logical polytope.

A projection of the 4-cube is retained.

Specific forms of the logical polytope have been applied to substantiate the steps of the drug discovery processes (Afshar et al. 2007, Luzeaux et al. 2008). The polytope describes in a general way a rational agent and enables the supervision of the computing process.

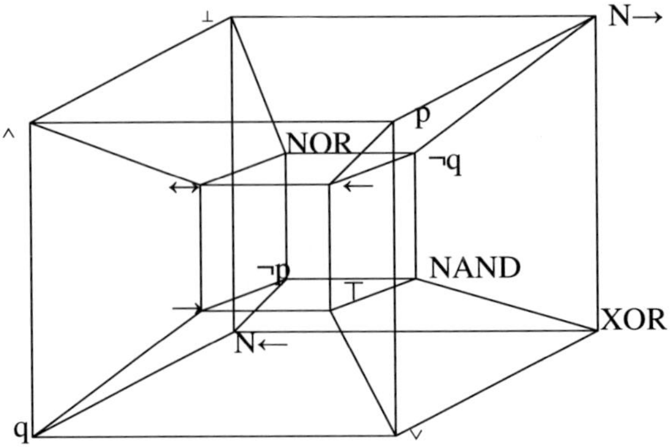

Fig. 7.3 Logical polytope

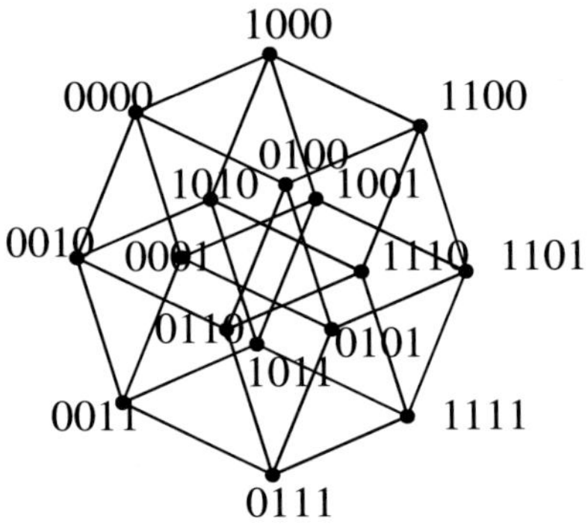

Fig. 7.4 Logical polytope sequence

7.3 Relational Complexity

A theory capable to analyze the processing demands of problems, to explain the main components of understanding and problem-solving methods was proposed by Halford (Halford 1993).

Structure mapping is the analogical reasoning that cognitive systems use to give meaning to problems by translating the given meaning of a problem into a representation or mental model that they already have and which allows them to understand the problem. The structure mappings that can be constructed depending upon the relational complexity of the structures they involve. The relational complexity of structures depends on the number of entities or the number of dimensions that are involved in the structure. The processing load of a task corresponds to the number of dimensions, which must be simultaneously represented, if their relations are to be understood. For example, to understand any comparison between two entities one must be able to represent two entities and one relation between them.

To understand a transitive relation, one must be able to represent at least three entities: otherwise it would not be possible to mentally arrange the entities in the right order that would reveal the relations between all entities involved.

Halford identified four levels of dimensionality for cognitive processes. The first is the level of element mappings. Mappings at this level are constructed on the basis of a single attribute. The second is the level of binary relations or relational mappings. At this level two-dimensional concepts can be constructed. Thus, two elements connected by a given relation can be considered at this level. The next is the level of system mappings, which requires that three elements or two relations must be considered simultaneously. At this level ternary relations or binary operations can be represented.

At the final level multiple-system mappings can be constructed. At this level quaternary relations or relations between binary operations can be constructed and four dimensions can be considered at once. The four levels of structure mappings correspond, in the theory of cognitive development of Piaget, to the sensorimotor, the preoperational, the concrete operational, and the formal stage. The four levels may be linked to the sensorimotor, interrelational, dimensional, and vectorial stages as described by Case (Case 1992).

In an overall sense there is a clear correspondence between Piaget's four major stages and the levels defined by Halford or by Case.

Fig. 7.5 shows the development stages-relational complexity polytope.

The elements of the front face of the polytope are presented in Table 7.2.

Table 7.2 outlines some categorification aspects for development stages.

Table 7.2 Categorification for development stages

Author\ Stage	K0	K1	K2	K3	Self
-	n=0	n=1	n=2	n=3	n≥4
Piaget (1971)	Sensori-motor	Preconceptual	Concrete Operational	Formal	Post-Formal
Halford (1993)	Elemental Association	Relational Mapping	Binary Operations	Quaternary Relations	-
Case (1992)	Sensori-motor	Interrelational	Dimensional	Vectorial	-

A challenge is the study of development stages for self-integrative closure, connecting levels n=0, sensory-motor and n=3, formal, and the emergence of the Self, corresponding to the levels n≥4 and to post-formal stages.

The notations for Fig. 7.5 are: K0-Elemental Association, K1-Relational Mapping, K2-Binary Operations, and K3-Quaternary Relation.

Fig. 7.5 outlines the direct integrative way S→K1→K2→K3 and the reverse way of differentiation K3′→K2′→K1′→S′.

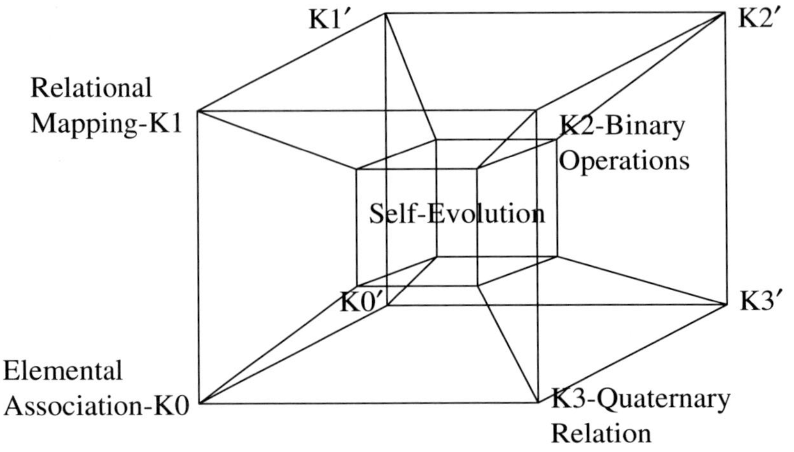

Fig. 7.5 Development stages: relational complexity polytope

A useful heuristic is that relational complexity cannot be reduced if the variables interact. This is analogous to analysis of variance method since interacting variables must be interpreted jointly. A procedure for determining effective relational complexity was described by Halford (Halford et al. 1998b). If a relation can be decomposed into simpler relations, then recomposed without loss of information, effective complexity is equivalent to the less complex relation.

The frontiers where new information is created consist of synchronized integrative and derivative ways. This explains why complex problem solving needs both integrative and derivative ways and the right rhythm of swinging between them.

The development stage theory of Piaget and the relational complexity theory open the problem of the level attained by different systems that learns and evolves.

The properties of higher cognitive processes and how they can be modeled by neural networks have been extensively studied by Halford and collaborators (Wilson and Halford 1994, Halford et al. 1998a, 1998b). They proposed and evaluated the so-called STAR (Structured Tensor Analogical Reasoning) model for problem solving.

The rank of tensor used in STAR is linked to the arity of relation, that is, to the number of attributes to the relation, and in the end, to the Piaget stages of cognitive development. The STAR model uses a tensor of rank-3 to represent a predicate of two arguments.

Halford studies suggest that for early Piaget stages in cognitive development, the categorical coproduct, "$\cup$", prevails allowing the associative knowledge. This is a fast and parallel process. During the higher Piaget stages the categorical product, "$\times$," seems preponderant, allowing the relational knowledge. It is a slow, sequential, effortful, higher cognitive process. The categorical product is naturally adapted to represent relations because its structure is analogous to the Cartesian product space in which relations are defined. The study of tensor product networks using distributed representations outlined the significant role of Hadamard matrices (Wilson and Halford 1994).

These matrices are special solutions of the wave equations.

The significance of Klein-4 group and of Latin squares for learning transfer in neural networks and in cognitive systems was also evaluated (Birney et al. 2006). Such structures are linked to the INRC group studied by Piaget (Inhelder and Piaget, 1958) as well as to standard solutions of the wave equation, WE model.

7.4 Explanatory Levels with n-Categories

Human inferential abilities like transitive inference and class inclusion, involve the dual category theory concepts, product and coproduct, respectively (Philips et al. 2009). Children around five years of age develop what is called transitive inference which is, for example, given that A is larger than B, and B is larger than C, one may infer that A is also larger than C. Class inclusion develops later in children and consists of the ability to discern between the cardinality of classes and subclasses.

Category theory shows that these abilities can be formally connected.

Transitive inference can be modeled with product, and class inclusion with its dual, the coproduct. This fact would explain that these two reasoning abilities have similar profiles of development, because they involve related sorts of processes, namely product and coproduct.

The n-category theory is useful to formally contrast category theory explanation against classical and connectionist approaches (Philips and Wilson 2010). Observe that the definitions of functor and natural transformation are very similar. In fact, they are morphisms at different levels of analysis. For n-category theory, a category such as Set is a 1-category, with 0-objects, that is, sets, for objects and 1-morphisms, that is, functions for arrows. A functor is morphism between categories. The category of categories, Cat, has categories for objects and functors for arrows. Thus, a functor is a 2-morphism between 1-objects, that is 1-categories, in a 2-category. A natural transformation is a morphism between functors. The functor category, Fun, has functors for objects and natural transformations for arrows. Thus, a natural transformation is a 3-morphism between 2-objects, that is functors, in a 3-category. A 0-category is just a discrete category, where the only arrows are identities, which are 0-morphisms. In this way, the order n of the category provides a formal notion of explanatory level (Phillips and Wilson 2010). Classical or connectionist compositionality is essentially a lower-level attempt to account for systematicity. That level is best described in terms of a 1-category. Indeed, a context-free grammar defined by a graph is modeled as the free category on that graph containing sets of terminal and non-terminal symbols for objects and productions for morphisms. By contrast, the category theory explanation involves higher levels of analysis, specifically functors and natural transformations, which live in 2-categories and 3-categories, respectively. Of course, one can also develop higher-order grammars that take as input or return as output other grammars. Similarly, one can develop higher-order networks that take as input or return as output other networks. The problem is that neither classical nor connectionist compositionality delineates those higher-order grammars or networks that have the systematicity property from those that do not.

Fig. 7.6 outlines the polytope for explanatory levels.

A decategorification way should be considered too. This means that after the integration way S→K1→K2→K3 we need to look at the differentiation way K3′→K2′→K1′→S′.

The differentiation is a kind of reverse epistemology. Observe that making use of the developments of the direct way, the reverse way may offer a symmetry-breaking results.

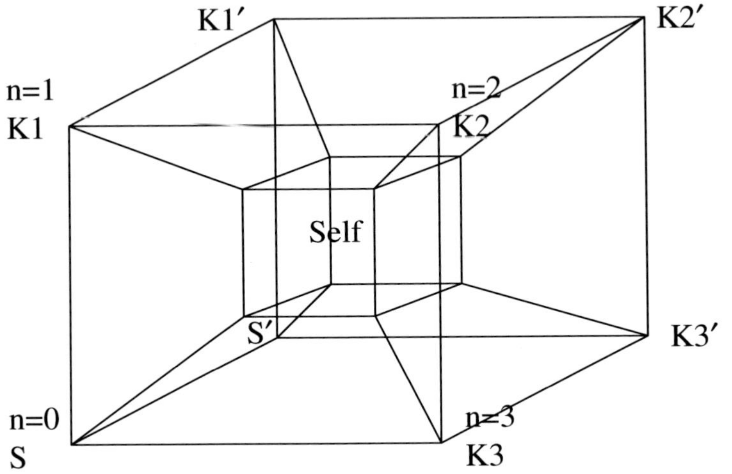

Fig. 7.6 Polytope for explanatory levels

In such cases, the swinging from direct to reverse epistemology is beneficial. Swinging methods based on direct and reverse epistemology have been applied in knowledge evaluation and development because the boundaries where new information is created consist of simultaneous tendencies. Tendencies to integrate should coexist with tendencies to differentiate and it is the intermixing of both that matters for self-evovability.

Table 7.3 outlines the categorification aspects for explanatory levels.

Table 7.3 Categorification for explanatory levels

Level	K0 (S)	K1	K2	K3	Self
-	n=0	n=1	n=2	n=3	n≥4
Categories	0-category	1-category	2-category	3-category	4-category
Example	sets	Set	Cat	Fun	-

The study of exploratory levels for self-integrative closure and the emergence of the Self corresponding to $n \geq 4$ are necessary.

7.5 LISA

LISA (Learning and Inference with Schemas and Analogies) is a system used in the synchronous activation approach to model analogical inference (Hummel and Holyoak 1997, Hummel and Choplin 2000). It demonstrates that temporal synchrony in conjunction with structured neural representations suffices to support complex forms of relational information processing specific to cognitive systems.

The problem for such systems is their suitability for reflexive or reflective cognitive processes. Reflexive processes are linked to categorical coproduct while reflective processes are linked to the categorical product. While reflexive and reflective processes follow different kinds of computational constraints, in most cases, the two types of processes interact and need to be integrated in the performance of a single task.

LISA is a computational model based on temporal synchrony and designed for analogical inference and for schemas induction.

LISA system is illustrated in Fig. 7.7. The basic level includes semantic units, s, the next includes the so-called localist units, L, (predicate/object or object/roles), the next level includes the sub-problems and the higher level the problems.

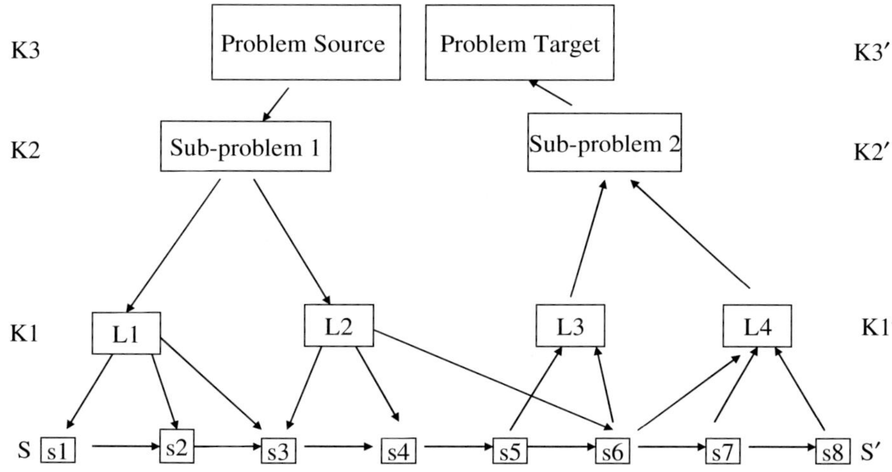

Fig. 7.7 LISA

LISA is a computational model based on temporal synchrony and designed for analogical inference and for schemas induction. The data for LISA network consists of a collection of trees and a representation that is a pattern of "0", "1" and so on for each terminal symbol occurring in those trees. The tree contains a hierarchy of entities: problem, sub-problems, roles, objects and semantics.

The task for the LISA network is to provide a means of compressing each tree into a representation, the so-called activation vector, and reconstructing the tree from its representation. The SKUP elements are naturally associated to the LISA elements. The problems to solve may be associated to the hierachy of conditions K1, K2 and K3. LISA contains a driver network associated to operators U, and to the reflective reasoning.

The representational structure of LISA provides at least a starting point for reflexive reasoning capabilities. LISA propositions are retrieved into memory via guided pattern matching. During retrieval and comparisons the proposition are divided into two mutually exclusive sets: a driver and one or more recipients or receivers. The receiver network is associated to possibilities P. Thc swinging between reflexive and reflective passes through the semantics. The LISA semantics elements are associated to the states S in SKUP.

The activation of semantic units is controlled by time. Often the analysts do not have the time to allow runaway activation of semantics since they needs make inferences quickly. Notice that in contrast to reflexive inferences which are fast, the reflective inferences may require more effort. An open problem is to establish, for imposed time frames, the number of swinging from reflexive to reflective and the order in which the swinging should be performed.

The Self takes into account the timescales for transition between levels. This allows the transition from problem source to problem target that is from integration way S→K1→K2→K3 and a differentiation way K3′→K2′→K1′→S′.

Inherently there appear differences between the two ways and this can be the source of creativity. That is because the boundaries where creative research grows require synchronized integration and differentiation tendencies.

Observe that this suppose that problem source and problem target are different.

Fig. 7.8 shows the polytope associated to LISA architecture.

The notations are: S-Semantic units, K1-Localist units, K2-Sub-problems, K3-Problems

Fig. 7.9 suggests a potential application of differential posets as cognitive architecture.

The D operator decomposes the problem while the U operator integrates and builds a problem target.

DORA (Discovery of Relations by Analogy) is a symbolic connectionist network that learns structured representations of relations from unstructured inputs. DORA is an extension of the LISA model of relational reasoning (Doumas et al. 2008).

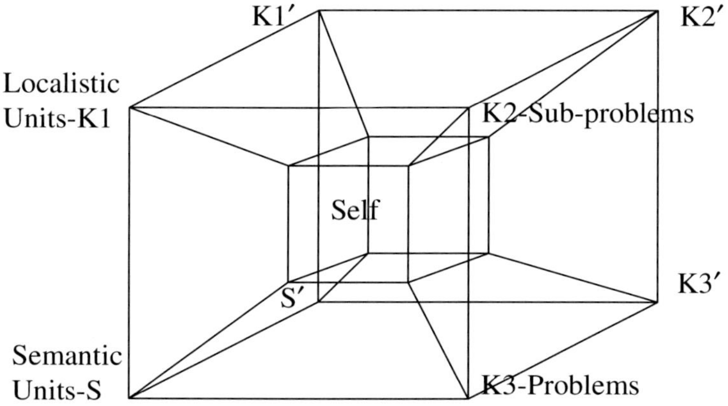

Fig. 7.8 Polytope for LISA framework

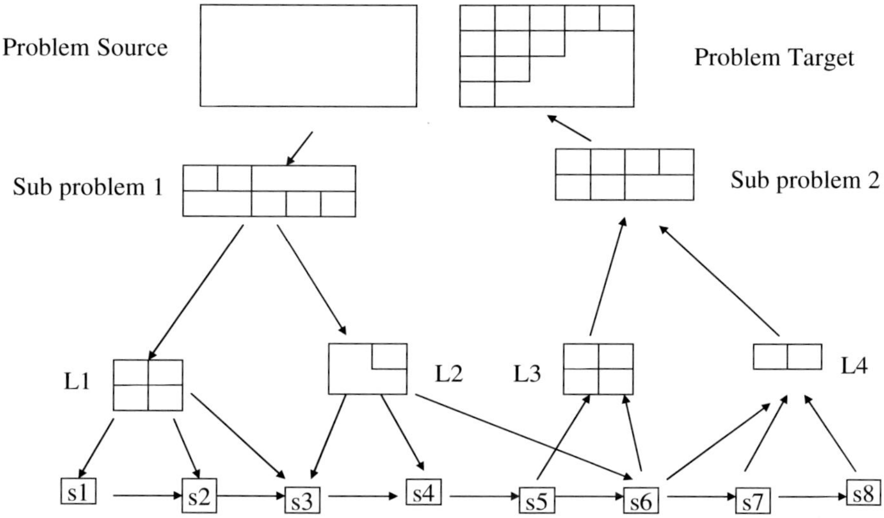

Fig. 7.9 Duality for LISA framework

DORA provides a means by which the representations used by LISA are learned from examples, and, consequently, provides an opportunity to understand the interplay between the dual sources of knowledge accumulation and increasing capacity limits as effectors of the changes in analogy making.

Like LISA, DORA dynamically binds distributed, that is connectionist, representations of relational roles and objects into explicitly relational, that is, symbolic, structures. The resulting representations enjoy the advantages of both connectionist and traditional symbolic approaches to knowledge representation, while suffering the limitations of neither. DORA's basic representational schema is adapted from LISA. In DORA, propositions are encoded by a hierarchy of structure unit.

Predicate and object units locally code for specific roles and fillers. While LISA must use different types of units to code for roles and their fillers, DORA uses the same types of units to code both roles and fillers and differentiates between roles and fillers via its binding mechanism. A comparison between DORA and STAR capabilities is due to Halford (Halford et al. 2010).

7.6 LIDA

LIDA (Learning Intelligent Distribution Agent) is a conceptual and computational framework for intelligent, autonomous, and conscious software agent that implements some ideas of the global workspace, GW, theory (Baars 2002).

LIDA appears as an attempt to adopt strategies observed in nature for creating information processing machinery.

The architecture is built upon the IDA (Intelligent Distribution Agent) framework, which was initially designed to automate the whole set of tasks of a human personnel agent who assigns resources to new tours of duty. LIDA employs a partly symbolic and partly connectionist memory organization, with all symbols being grounded in the physical world (Franklin 2006, Baars and Franklin 2009).

Baars' GW theory has inspired a variety of related consciousness models (Baars 1988). The central idea of GW theory is that conscious cognitive content is globally available for diverse cognitive processes including attention, evaluation, memory, and verbal report. The notion of global availability is suggested to explain the association of consciousness with integrative cognitive processes such as attention, decision making and action selection. Also, because global availability is necessarily limited to a single stream of content, GW theory may naturally account for the serial nature of conscious experience.

GW theory was originally described in terms of a blackboard architecture in which separate, quasi-independent processing modules interface with a centralized, globally available resource. This cognitive level of description is preserved in the computational models of Franklin, who proposed a model consisting of a population of interacting software agents, and Shanahan, whose model incorporates aspects of internal simulation supporting executive control and more recently spiking neurons (Shanahan 2006, 2008).

A central global workspace, GW, constituted by long-range cortico-cortical connections, assimilates other processes according to their salience. Other automatically activated processors do not enter the global workspace.

A neuronal implementation of a global workspace, GW, architecture, the so-called neuronal global workspace was studied (Dehaene et al. 2003).

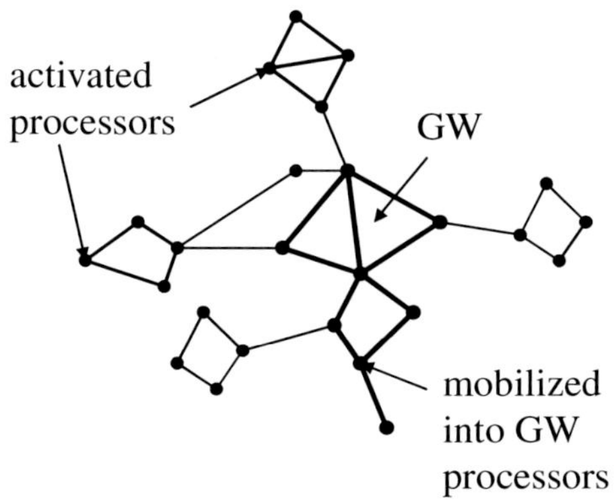

Fig. 7.10 Diagram for neuronal global workspace

Fig. 7.10 contains a schematic of the neuronal global workspace.

In this model, sensory stimuli mobilize excitatory neurons with long-range cortico-cortical axons, leading to the genesis of a global activity pattern among workspace neurons. Any such global pattern can inhibit alternative activity patterns among workspace neurons, thus preventing the conscious processing of alternative stimuli, for example, during the so-called attentional blink. The global neuronal workspace model predicts that conscious presence is a nonlinear function of stimulus salience; that is, a gradual increase in stimulus visibility should be accompanied by a sudden transition of the neuronal workspace into a corresponding activity pattern (Dehaene et al. 2003).

The complementary role of the conscious and unconscious for cognition and self-evolvability was emphasized.

The swinging between conscious and unconscious is an important tool for designing creative systems that can autonomously find solutions to highly complex and ill-defined construction problems.

When a module pl invades the workspace, the others, as p2 are blocked at a similar depth.

Fig 7.11 illustrates the global workspace architecture activity.

In GW theory the processes, p1, p2 and so on, said to be unconscious, compete to enter the global workspace GW. This competition is at several levels.

Such processes are often thought of as memory activities, as for instance episodic or working memories.

Suppose that there are two levels of competition indexed by K1 and K2 and the competition is won by one process, for instance p2.

Having entered the GW, the winning process becomes the conscious state of the system. This is continuously broadcast back to the originating processes that change their state according to the conscious state. This results in a new conscious state and so on linking sensory input to memory and conscious states.

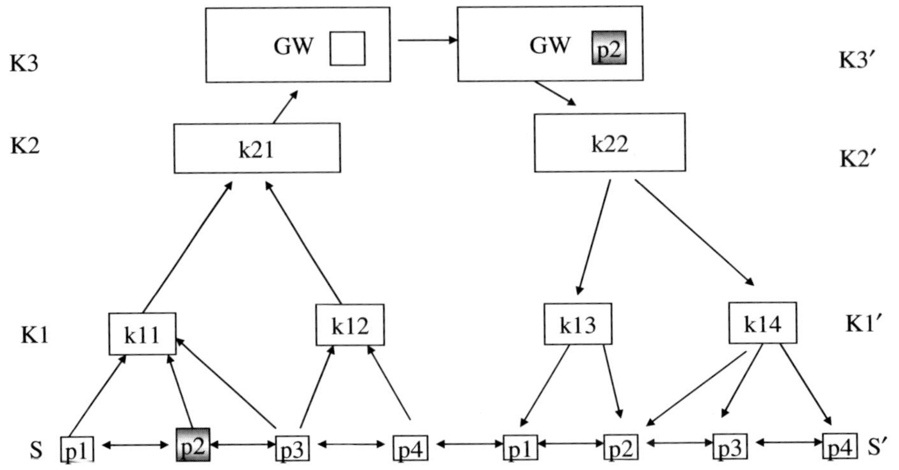

Fig. 7.11 Global workspace architecture

LIDA has distinct modules for perception, working memory, semantic memory, episodic memory, action selection, expectation and automatization (learning procedural tasks from experience), constraint satisfaction, deliberation, negotiation, problem solving, metacognition, and conscious-like behavior. Most operations are done by codelets implementing the unconscious processors, that is, specialized networks of the global workspace theory. A codelet is a small piece of code or program that performs one specialized, simple task. The LIDA framework incorporates three new modes of learning into the older IDA model: perceptual, episodic, and procedural learning, which are all of bottom-up type. Perceptual learning concerns learning of new objects, categories, relations, and so on, and takes two forms: strengthening or weakening of the base-level activation of nodes, as well as creation of new nodes and links in the perceptual memory. Episodic learning, on the other hand, involves learning to memorize specific events that is, the what, where, and when. It results from events taken from the content of *consciousness* being encoded in the transient episodic memory. Finally, procedural learning concerns learning of new actions and action sequences with which to accomplish new tasks. This combines selectionist learning that is, selecting from an obsolete repertoire, and the instructionalist learning, that is, constructing new representations, with functional consciousness providing reinforcements to actions. This architecture may explain many features of mind, however, it remains to be see whether high competence will be achieved in understanding language, vision, and common sense reasoning based on perceptions.

The LIDA model covers a large portion of human-like cognition (Franklin and Patterson 2006). Based primarily on GW theory the model implements a number of psychological and neuropsychological theories.

The LIDA computational architecture is derived from the LIDA cognitive model. The LIDA model and its ensuing architecture are grounded in the LIDA cognitive cycle. Every autonomous agent, human, animal, or artificial, must frequently sample and sense its environment and select an appropriate response, an action.

More sophisticated agents, such as humans, processes make sense of the input from such sampling in order to facilitate their decision making. The agent's life can be viewed as consisting of a continual sequence of these cognitive cycles. Each cycle constitutes a unit of sensing, attending and acting.

A cognitive cycle can be thought of as a moment of cognition, a cognitive moment.

During each cognitive cycle the LIDA agent first makes sense of its current situation as best as it can by updating its representation of its current situation, both external and internal. By a competitive process, as specified by GW theory, it then decides what portion of the represented situation is most in need of attention. Broadcasting this portion, the current contents of consciousness enable the agent to choose an appropriate action and execute it, completing the cycle.

Thus, the LIDA cognitive cycle can be subdivided into three phases, the understanding phase, the attention that is, the consciousness phase, and the action selection phase. Fig. 7.12 illustrates some elements of LIDA architecture. It starts in the lower-left corner and develops roughly clockwise (Snaider et al. 2011).

The first module is denoted by S. During the understanding phase, incoming stimuli activate low-level feature detectors in Sensory Memory. The output is sent to Perceptual Associative Memory, where higher-level feature detectors feed in to more abstract entities such as objects, categories, actions, events, and so on. The resulting percept moves to the Workspace, denoted by K1. Here it triggers both Transient Episodic Memory, and Declarative Memory, producing local associations. These local associations are combined with the percept to generate a Current Situational Model, which represents the agent's understanding of what is going on right now.

Attention Codelets, associated here by K2, begins the attention phase by forming coalitions of selected portions of the Current Situational Model and moving them to the GW.

A competition in the GW then selects the most salient, the most relevant, the most important, and the most urgent coalition whose contents become the content of consciousness. These conscious contents are then broadcast globally, initiating the action selection phase, associated here to K3.

The GW space corresponds to the Self. The neuronal global workspace, GW appears in Fig. 7.12 as a working example of the Self.

The action selection phase of LIDA's cognitive cycle is also a learning phase in which several processes operate in parallel.

New entities and associations, and the reinforcement of old ones, occur as the conscious broadcast reaches Perceptual Associative Memory. Events from the conscious broadcast are encoded as new memories in Transient Episodic Memory.

Possible action schemas, together with their contexts and expected results, are learned into Procedural Memory from the conscious broadcast. Older schemas are reinforced.

In parallel with all this learning, and using the conscious contents, possible action schemas are recruited from Procedural Memory. A copy of each such schema is instantiated with its variables bound and sent to Action Selection, where it competes to be the behavior selected for this cognitive cycle. The selected behavior triggers Sensory-Motor Memory to produce a suitable algorithm for the execution of the behavior.

Its execution completes the cognitive cycle.

The Workspace requires further explanation. Its internal structure is composed of various input buffers and three main modules: the Current Situational Model, the Scratchpad and the Conscious Contents Queue. The Current Situational Model is where the structures representing the actual current internal and external events are stored. Structure-building codelets are responsible for the creation of these structures using elements from the various submodules of the Workspace. The Scratchpad is an auxiliary space in the Workspace where structure-building codelets can construct possible structures prior to moving them to the Current Situational Model. The Conscious Contents Queue holds the contents of the last several broadcasts and permits LIDA to understand and manipulate time-related concepts.

The GW mediates between the direct integrative way S→K1→K2→K3 and the reverse differentiation way K3′→K2′→K1′→S′ as shown by Fig. 7.11 and fig. 712.

The reverse epistemology allows making use of the developments of the direct way and will offer is a kind of symmetry-breaking result. The swinging from direct to reverse epistemology is beneficial.

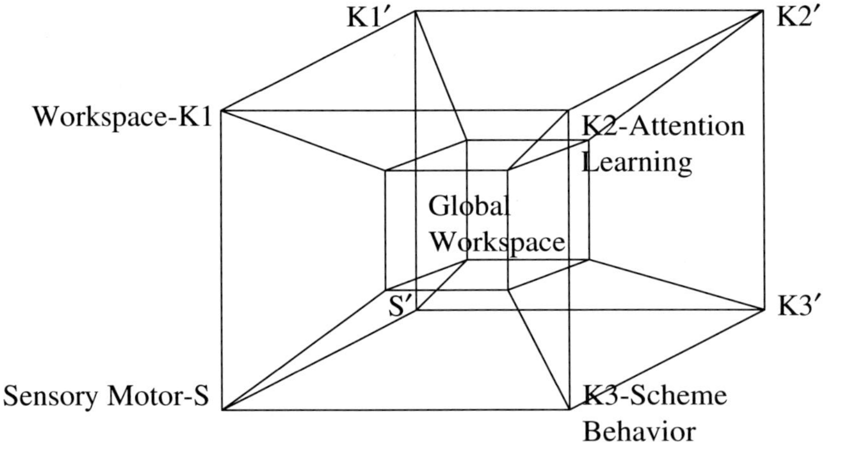

Fig. 7.12 Polytope for LIDA framework

The wave character manifested as swinging behavior is applied for evaluation and creative behavior. The boundaries where creative research grows and new information is created consist of synchronized tendencies. Tendencies to integrate should coexist with tendencies to differentiate and it is the blend of both that counts for self-evolvability.

Fig. 7.13 suggests a potential application of differential posets as cognitive architecture.

The U operator transfer processes as p2 to the GW space while the D operator transfer processes from GW toward field.

One feature of human thought not accounted for by the GW theory is the reflexivity.

This is the capacity for a conscious thought to refer to itself or to other conscious states.

Consider that thought is internally in simulation with the environment. This simulation hypothesis can explain our experience of an inner world.

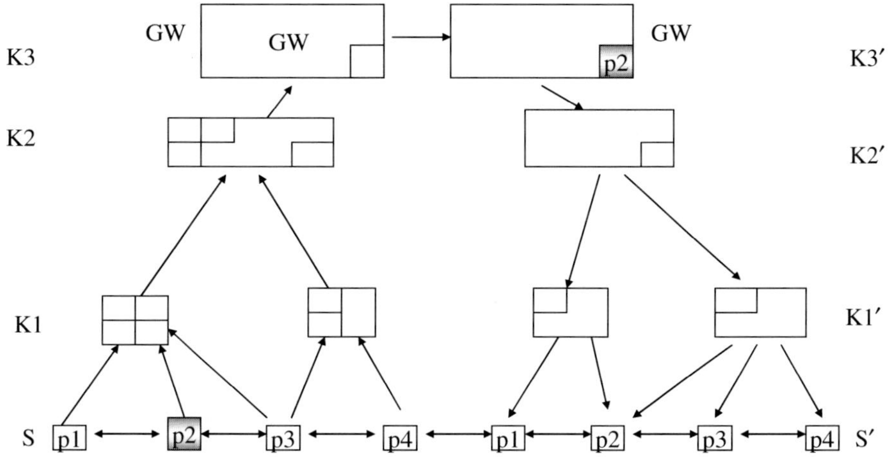

Fig. 7.13 Duality for LIDA framework

The simulation hypothesis is based on the following assertions:

• The brain's motor centers can be active without producing explicit action

• The brain's perceptual apparatus can be active without the presence of external stimuli

• Internally generated motor activity can elicit internally generated perceptual activity through associative mechanisms

By augmenting the basic GW workspace architecture with an internally closed loop it is possible to reconcile the GW theory with the so-called simulation hypothesis (Shanahan 2006). The proposal is in support of the hypothesis that organisms whose brains are endowed with such an internal loop are capable of rehearsing the consequences of potential actions prior to actually carrying them out.

Such implementations are useful as a proof-of-concept, despite the present lack of neurological plausibility, both at the level of the neuron model used and in its employment of a single attractor network to model the global workspace.

Finally it should be observed that LISA, LIDA and global workspace GW theory are similar approaches. They mix serial and parallel computations, corresponding to different types of categorical product.

This supports their study by similar polytopic architectures.

7.7 *Physarum* Computing Systems

The slime mold *Physarum polycephalum* is a multinuclear, single-celled organism that has properties making it ideal for the study of resource distribution networks and of cognitive capabilities (Nakagaki et al. 2000, Nakagaki 2001).

The organism is a single cell, but it can grow to tens of centimeters in size so that it can be studied and manipulated with modest laboratory facilities.

The presence of nutrients in the cell body triggers a sequence of chemical reactions leading to oscillations along the cell body. Tubes self-assemble perpendicular to the oscillatory waves to create networks linking nutrient sources throughout the cell body. There are two key mechanisms in the slime mold life cycle that transfer readily to resource distribution network problems. First, during the growth cycle, the slime mold explores its immediate surroundings with pseudopodia via chemotaxis to discover new food sources. The second key mechanism is the temporal evolution of existing routes through nonlinear feedback to efficiently distribute nutrients throughout the organism. In slime mold, it can be shown experimentally that the diameters of tubes carrying large fluxes of nutrients grow to expand their capacity, and tubes that are not used decline and can disappear entirely. Unlike any other circulatory system, networks in slime mold rebuild themselves dynamically to changing environmental conditions.

Nakagaki proposed a simple yet powerful model for tube evolution in *Physarum* to reproduce slime mold maze-solving experiments (Nakagaki 2001). The model captures the evolution tube capacities in an existing network through a coupled system of ordinary differential equations. Flow through the network is driven by a pressure at each node. The diameter of the tubes evolves based on the flux of nutrients through the network.

Nakagaki's group makes considerable claims about robustness and intelligence level in the *Physarum* colonies (Nakagaki et al. 2000, 2001, 2004).

Implementation of a general-purpose computing machine is the most remarkable feature of the plasmodium of *Physarum*.

The cognitive levels of *Physarum* may be compared to these attained by some pointer machines (Ben-Amram 1995, 1998).

Experimentally it was demonstrated that the plasmodium can implement the Kolmogorov–Uspensky Machine (KUM), a mathematical machine in which the storage structure is an irregular graph (Adamatzky 2007). The KUM is a forerunner and direct ancestor of Schoenhage's storage modification machines (Schoenhage 1980).The storage modification machines are basic architectures for random access machines, which represent the basic architecture of modern-day computers. The plasmodium-based implementation of KUM provides a biological prototype of a general-purpose computer.

The key component of the KUM is an active zone, which may be seen as a computational equivalent to the head in a Turing machine. Physical control of the active zone is of utmost importance because it determines functionality of the biological storage modification machine.

Laboratory and computer experiments with *Physarum* show basic operations Add node, Add edge, Remove Edge implemented in the *Physarum* machine. They also provide results on controlling movement of an active zone.

The filaments movements for *Physarum* suggest that their capabilities are at the level of the 1-categories and 2-categories. For 2-categories the pentagon relation is valid.

Physarum is able to disconnect a filament and reconnect it in another position.

This corresponds to 2-categories. At operadic level this corresponds to the associahedron K_4.

For 3-categories the so-called pentagon of pentagons or the associahedron K_5 should be considered. This needs a spatial awareness that allows evaluating the *Physarum* computing capabilities.

Table 7.4 shows the knowledge level associated to different associahedra

It refers to categorification aspects.

Needed are the study of exploratory levels for self-integrative closure and the emergence of the Self corresponding to n≥4.

An interesting test for *Physarum* capabilities would be the evolution in a high-dimensional space with restrictions as shown in Fig. 7.14.

Table 7.4 Categorification for associahedra

Level	K0	K1	K2	K3	Self
	n=0	n=1	n=2	N=3	n≥4
Categories	0-category	1-category	2-category	3-category	4-category
Associahedra	K(2)	K(3)	K(4)	K(5)	K(6)
Geometry	-	Trees	Pentagon	Pentagon of Pentagons	-

Figure 7.14 illustrates the node-disjoint path construction between the source $x = 0000$ and the destination $y = 1110$ in a 4-cube.

The edges on the four node-disjoint paths are labeled with the corresponding dimensions. Since x and y differ in bits 0, 1, and 2, the four paths correspond to the dimension sequences (0,1,2), (1,2,0), (2,0,1), and (3,0,1,2,3). At least one of these paths is fault-free in the presence of any three or less faulty nodes.

Tsuda described an experimental setup that interfaces an amoeboid plasmodium of *Physarum* with an omni-directional hexapod robot to realize an interaction loop between environment and plasticity in control (Tsuda et al. 2006). Through this bio-electronic hybrid architecture the continuous negotiation process between local intracellular reconfiguration on the micro-physical scale and global behavior of the cell in a macroscale environment can be studied in a device setting.

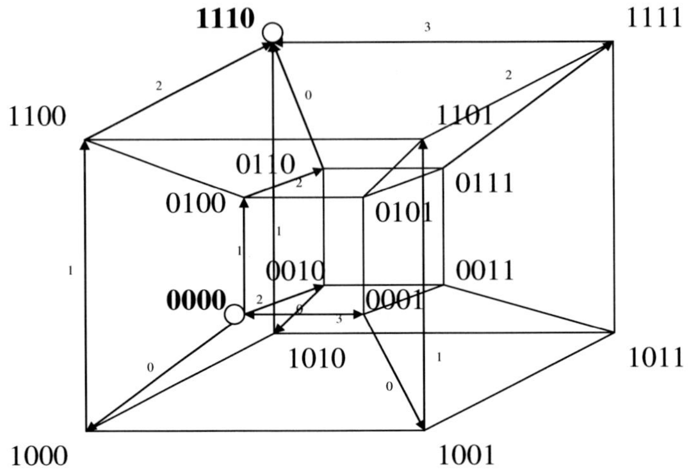

Fig. 7.14 Disjoint paths in 4-cube

The filaments movements for *Physarum* suggest that their capabilities are at the level of 1-categories and 2-categories. For 2-categories the pentagon relation is valid. This means cognitive capabilities

For 3-categories the pentagon of pentagons or a kind of spatial sensitivity should be considered.

References

Adamatzky, A.: Physarum machine: implementation of a Kolmogorov-Uspensky machine on a biological substrate. Parallel Processing Letters 17, 455–467 (2007)

Afshar, M., Dartnell, C., Luzeaux, D., Sallantin, J., Tognetti, Y.: Aristotle's Square Revisited to Frame Discovery Science. Journal of Computers 2(5), 54–66 (2007)

Baars, B.J.: A cognitive theory of consciousness. Cambridge University Press, Cambridge (1988)

Baars, B.J.: The conscious access hypothesis: origins and recent evidence. Trends in Cognitive Science 6, 47–52 (2002)

Baars, B.J., Franklin, S.: Consciousness is computational: The LIDA model of Global Workspace Theory. International Journal of Machine Consciousness 1(1), 23–32 (2009)

Ben-Amram, A.M.: What is a Pointer machine? SIGACT News, ACM Special Interest Group on Automata and Computability Theory 26 (1995)

Ben-Amran, A.M.: Pointer machines and pointer algorithms: an annotated bibliography (1998), `http://www2.mta.ac.il/amirben`

Birney, D.P., Halford, G.S., Andrews, G.: Measuring the influence of complexity on relational reasoning: The development of the Latin Square Task. Educational and Psychological Measurement 66, 146–171 (2006)

Bringsjord, S., Taylor, J., Wojtowicz, R., Arkoudas, K., van Heuvlen, B.: Piagetian Roboethics via Category Theory: Moving Beyond Mere Formal Operations to Engineer Robots Whose Decisions are Guaranteed to be Ethically Correct. In: Anderson, M., Anderson, S. (eds.) Machine Ethics. Cambridge University Press, Cambridge (2010)

Case, R.: The mind's staircase: Exploring the conceptual underpinnings of children's thought and knowledge. Erlbaum, Hillsdale (1992)

Dehaene, S., Sergent, C., Changeux, J.P.: A neuronal network model linking subjective reports and objective physiological data during conscious perception. Proc. Natl. Acad. Sci. USA 100, 8520–8525 (2003)

Doumas, L.A.A., Hummel, J.E., Sandhofer, C.M.: A theory of the discovery and predication of relational concepts. Psychological Review 115, 1–43 (2008)

Franklin, S.: The LIDA architecture: Adding new modes of learning to an intelligent, autonomous, software agent. In: Proc. of the Int. Conf. on Integrated Design and Process Technology. Society for Design and Process Science, San Diego (2006)

Franklin, S., Patterson, F.G.J.: The LIDA Architecture: Adding New Modes of Learning to an Intelligent, Autonomous, Software Agent. In: IDPT 2006 Proceedings Integrated Design and Process Technology. Society for Design and Process Science (2006)

Halford, G.S.: Children's understanding: The development of mental models. Erlbaum, Hillsdale (1993)

Halford, G.S., Wilson, W.H., Phillips, S.: Processing capacity defined by relational complexity. Implications for comparative, developmental and cognitive psychology. Behavioural and Brain Sciences 21(6), 803–831 (1998a)

Halford, G.S., Wilson, W.H., Phillips, W.: Relational complexity metric is effective when assessments are based on actual cognitive processes. Behavioral Brain Sciences 21(6), 803–864 (1998b)

Halford, G.S., Wilson, W.H., Phillips, S.: Relational knowledge: the foundation of higher cognition. Trends in Cognitive Science 14(11), 497–505 (2010)

Hummel, J.E., Choplin, J.M.: Toward an integrated account of reflexive and reflective reasoning. In: Gleitman, L.R., Joshi, A.K. (eds.) Proceedings of the Twenty Second Annual Conference of the Cognitive Science Society, pp. 232–237. LEA, Mahwah (2000)

Hummel, J.E., Holyoak, K.J.: Distributed representation of structure. A theory of analogical access and mapping. Psychological Review 104, 427–466 (1997)

Inhelder, B., Piaget, J.: The Growth of Logical Thinking from Childhood to Adolescence. Basic Books, New York (1958)

Luzeaux, D., Sallantin, J., Dartnell, C.: Logical extensions of Aristotle's square. Logica Universalis 2, 167–187 (2008)

Moretti, A.: The Geometry of Logical Opposition. PhD. Thesis. University of Neuchâtel, Switzerland (2009)

Nakagaki, T.: Smart behavior of true slime mold in a labyrinth. Research in Microbiology 152(9), 767–770 (2001)

Nakagaki, T., Yamada, H., Hara, M.: Smart network solutions in an amoeboid organism. Biophysical Chemistry 107(1), 1–5 (2004)
Nakagaki, T., Yamada, H., Toth, A.: Maze-solving by an amoeboid organizm. Nature 407(6803), 470–470 (2000)
Nakagaki, T., Yamada, H., Toth, A.: Path finding by tube morphogenesis in an amoeboid organism. Biophysical Chemistry 92(1-2), 47–52 (2001)
Philips, S., Wilson, W.H., Hadford, G.: What do transitive inference and class inclusion have in common? categorical (co)products and cognitive development. PLoS Comput. Biol. 5(12), e1000599 (2009)
Philips, S., Wilson, W.H.: Categorial compositionality: A category theory explanation for the systematicity of human cognition. PLoS Comput. Biol. 6(7), 1–14 (2010)
Piaget, J.: Genetic Epistemology. Columbia University Press, New York (1970)
Piaget, J.: The construction of Reality in the Child. Ballantine Books, New York (1971)
Piaget, J.: Introduction a l'epistemologie genetique. La pensee Mathematique. Presses Universitaires de France, Paris (1973)
Schoenhage, A.: Storage modification machines. SIAM J. Comp. 9, 490–508 (1980)
Shanahan, M.: A cognitive architecture that combines internal simulation with a global workspace. Consciousness and Cognition 15, 433–449 (2006)
Shanahan, M.: A spiking neuron model of cortical broadcast and competition. Conscious Cogn. 17(1), 288–303 (2008)
Snaider, J., McCall, R., Franklin, S.: The LIDA Framework as a General Tool for AGI. In: The Fourth Conference on Artificial General Intelligence, Mountain View, CA (2011)
Tsuda, S., Zauner, K.-P., Gunji, Y.-P.: Robot Control: From Silicon Circuitry to Cells. In: Ijspeert, A.J., Masuzawa, T., Kusumoto, S. (eds.) BioADIT 2006. LNCS, vol. 3853, pp. 20–32. Springer, Heidelberg (2006)
Wilson, H.W., Halford, G.S.: Robustness of Tensor Product networks using distributed representations. In: Proceedings of the Fifth Australian Conference on Neural Networks (ACNN 1994), pp. 258–261 (1994)
Yang, Y., Bringsjord, S.: Mental Metalogic: A New, Unifying Theory of Human and Machine Reasoning. Erlbaum, Mahwah (2005)

Chapter 8
Control Systems

Abstract. Self-evolvability aspects for autonomous robots and high dimensional automata are illustrated.

Self-configuring schemas for self-evolvable control are correlated to general PSM frameworks. Entropy criteria prove to be useful to evaluate control architectures.

Different types of interconnections are reviewed and associated to wave equation, WE, solutions.

8.1 Self-Evolvable Control Systems

Trends in complex, software-intensive control systems show a continuous process of incorporating mechanisms of self-representation and self-reflection. These mechanisms allow improving system performance and resilience in changing uncertain environments. Systems reflect upon themselves by means of self-models. From plant representations like those employed in classic model-based adaptive controllers to encompassing plant/controller mixed models, the internal model principle is pushing software-intensive control systems designs to a scale that would match the complexity of models of human-like self-awareness and consciousness (Sanz et al. 2005).

The domain of automatic control deeply trapped in the limited mathematics of linear systems has not rendered the promised artificial intelligence. Cognitive architecture is still far from offering the minimal glimpse of a human mind. Only few human-like capabilities were sought to improve localized control systems performance.

For different objectives, as humanoid automata or intelligent controllers, there is a need of going beyond what we are able to do today and search the seemingly missing essence of mind.

The search for the essence of mind has been a major pursuit in different fields-robotics, neuroscience, psychology mathematics, and philosophy that have converged into a programmatic discipline, the cognitive science.

This unified vision is powerful and is providing a way for trying to formalize challenges as perception, knowledge, thought or even consciousness.

The intelligent control tried to mimic concrete human thought processes in search for competence.

The research leads to the conclusion that one viable strategy to eliminate brittleness and increase mission-level resilience is to make systems epistemologically robust at the mission level, so we can move the responsibility for real-time cognitive behavior from engineers to the systems themselves during runtime. And to do this we need the self-consciousness.

One of the critical elements in this approach is the epistemic control loop.

Even if there are arguments against the possibility of machine self-awareness several attempts at realizations have been done recently (Sanz et al. 2007, 2011).

Fig. 8.1 shows the abstract architecture for a self-aware, conscious control.

There are two control loops to be considered.

The external loop includes S-Substrate, K1-Sense Perceive, K2-Model Control, and K3-Act.

This corresponds to the epistemic control loop, a primary form of perception before the Self, in other words, the conscious level. This level appears as prototype of the Self.

The internal loop includes the tips, s-substrate, k1-sense, perceive, k2-meta-model, meta-control, and k3-reconfiguration.

A self-aware system is continuously generating meanings from continuously up-date self-models. The agent perceives and controls itself as it perceives and controls the world.

The Self allows self-integrative closure and may make the entire system self-evolvable and fully autonomous.

This suggests that after the integration way K0→K1→K2→K3 we need to look at the differentiation way K3′→K2′→K1′→K0′. The swinging from direct to reverse investigation is critical for model evaluation and self-evolution.

Since the boundaries where creative research grows consist of coexisting tendencies, integration should coexist with tendencies to differentiation and it is the timing and intermixing of both that counts.

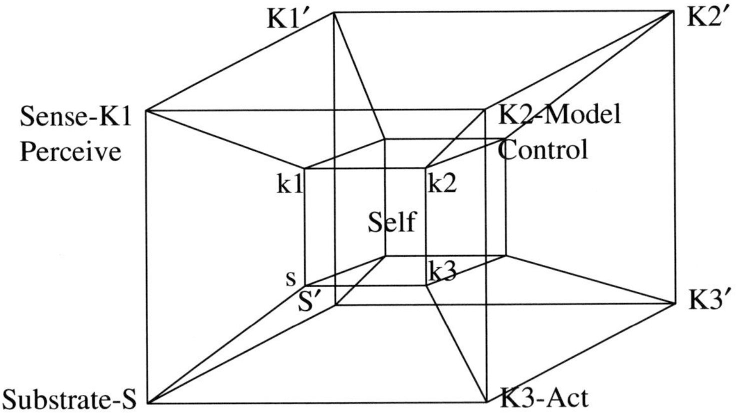

Fig. 8.1 Polytope for self-evolvable control

A robust autonomous system as shown in Fig. 8.1 will not only realize a hierarchical federation of cognitive control loops but also a transversal metacognitive capability mediated by the Self. This will render the necessary self-awareness for achieving full autonomy.

8.2 Self-Evolvable Robots

Autonomous robots should evolve in non-cooperative even hostile outdoor environments. Reaction to disturbance is a first step toward autonomy.

A more demanding definition of autonomy includes the ability to change the interaction modes with the environment. An autonomous organization has to internalize external constraints, which means the ability to integrate knowledge of its own dynamics and representation of the exterior. Such ability is closely connected to the self-awareness of a frontier between the inside and outside of the system, which means operational closure (Maturana and Varela 1992). A general overview of control architectures for robots is due to Arkin (Arkin 1998).

Fig. 8.2 shows the autonomous robots architecture (Luzeaux et al. 2001).

For this autonomous robots architecture, the levels may be identified as: K0-Perception, K1-Attention Manager, K2-Behavior Selection, and K3-Action. To these the central meta-representation linked to the previous levels is joined.

The center may be considered as the final target of one cycle of environment investigations.

Sensors yield data to perception algorithms which create representations of the environment. These perception processes are activated or inhibited by the attention manager and receive also information on the current executed behavior. This information is used to check the consistency of the representation. The attention manager periodically updates representations. The action selection module chooses the robots behavior depending on the predefined goals, the current action, the representations and their reliability.

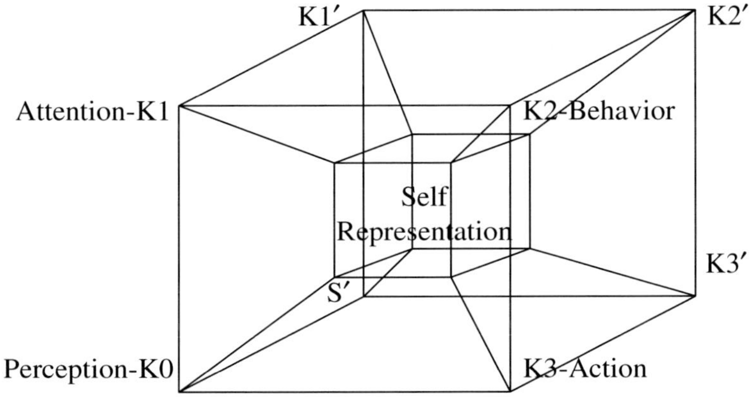

Fig. 8.2 Polytope for self-evolvable robots

Finally the behaviors control the robots actuators in closed loop with the associated perception processes.

This modular architecture allows developing independently the various processes belonging to each of the four basic entities, integrating them together.

The Self is considered either as the starting area or as the final area of one cycle of investigations. The periodic swinging between the two roles should be considered too.

This suggests that after the integration way K0→K1→K2→K3 we need to look at the differentiation way K3′→K2′→K1′→K0′. The swinging from direct to reverse investigation would be beneficial for model evaluation.

Current developments focus on the extension of architecture to multiple robots. A major issue is to determine how the representation of every robot can be shared and how individual behavior selection can take the other robots' selected behavior into account, allowing group missions such as scouting.

Fig. 8.3 illustrates the operad for self-evolvable robots.

The notations are: K0-CPA-communication perception agent, K1-AMA-attention manager agent, K2-BSA-behavior selection agent, K3-ASA-action selection agent.

Here the Self indicates the way to put together, while the four surrounding modules indicate things to put together.

A capability to act successfully in a complex, ambiguous, and harsh environment would vastly increase the application domain of robotic devices. Established methods for robot control run up against a complexity barrier. The living organisms amply demonstrate that this barrier is not a fundamental limitation.

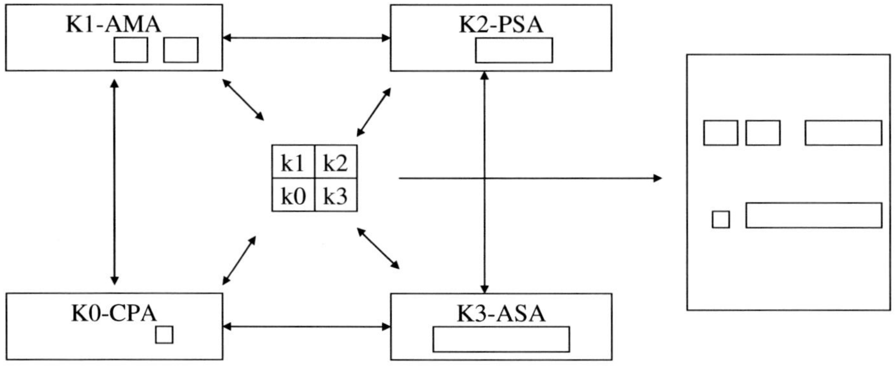

Fig. 8.3 Operad for self-evolvable robots

8.3 Self-Functioning Systems

Sensor processing supplies information to the knowledge representation repository, and the detection of any change in state triggers behavior generation. The behavior generator, in attempt to check whether new information may suggest a deviation from the plan or a potential threat, will initiate a simulation loop. The intelligent behavior emerges when the behavior generator implicitly raises several questions and attempts to simulate a number of solutions. This thinking loop is the Elementary Loop of Self- Functioning, ELSF shown in Figure 8.4.

A goal or sets of goals together with the ability to detect changes within the world model allow for the actual decision making. The ELSF becomes a critical building block in the Intelligent Node infrastructure (Dawidowicz 1999, 2002).

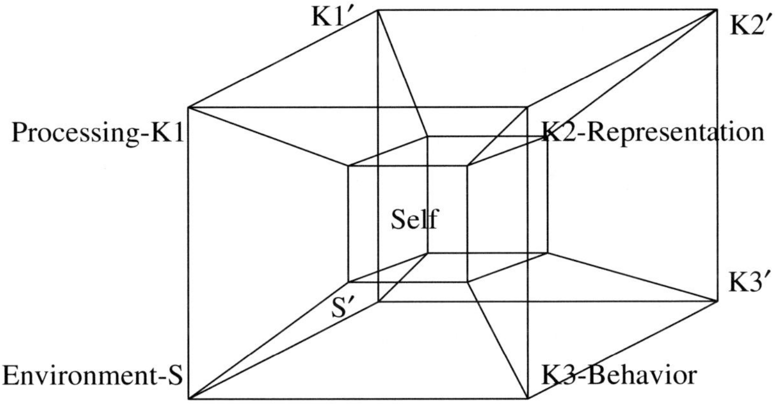

Fig. 8.4 Polytope for elementary loop of self-functioning

Fig. 8.4 shows the modules of the elementary loop of self-functioning.

The notations are: S-Environment, K1-Processing, K2-Representation, and K3-Behavior

The Self allows self-integrative closure and may make the entire system self-evolvable.

Fig. 8.5 shows the elementary loop of self-functioning with multi-resolution.

The multi-resolution case is explained in what follows (Dawidowicz 1999, 2002).

Each time we focus our attention on the object's details or the smaller objects that make up the object of our attention, we go down a level to a level of higher resolution. The level of resolution is proportional to the degree of detail required to describe an object. The level of resolution is relative. The upper echelons make decisions in low levels of resolution and require more abstract concepts to make decisions. The lower echelons receive orders from upper echelons and have to interpret them using a higher level of resolution. For more effective communication the ideas are expressed at the lowest possible resolution. The ideas and concepts emerge or come to the low-level surface as a result of generalizations at a higher level of resolution. The upper echelon requires a low level of resolution while the low echelon requires a high-resolution level of resolution.

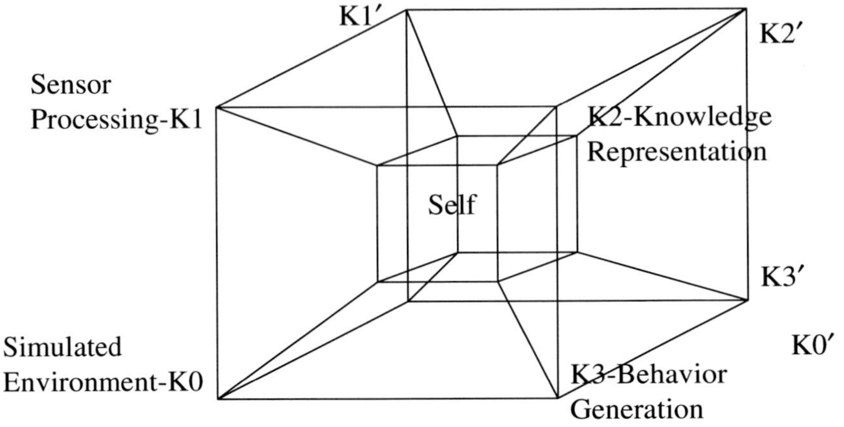

Fig. 8.5 Elementary loop of self-functioning with multi-resolution

The notations in Fig. 8.5 are: K0-Simulated Environment, K1-Sensor Processing, K2-Knowledge Representation, and K3-Behavior Generation.

Fig. 8.4 and Fig. 8.5 suggest that after the integration way K0→K1→K2→K3 we need to look at the differentiation way K3′→K2′→K1′→K0′. The swinging from direct to reverse way counts for self-evolvability implementation.

Fig. 8.6 shows the operad associated to self-functioning.

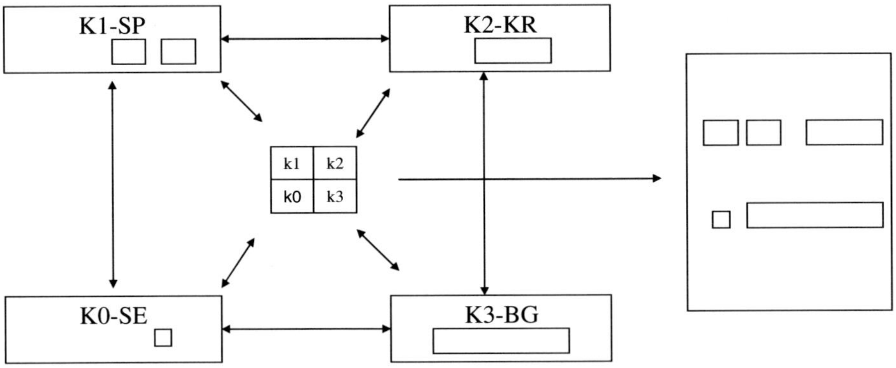

Fig. 8.6 Operad for self-functioning systems

The notations for Fig. 8.6 are: K0-Simulated Environment, SE, K1-Sensor Processing, SP, K2-Knowledge Representation, KR, and K3-Behavior Generation, BG.

The Self indicates the way to put together, while the four surrounding modules indicate things to put together.

In this case k0 activates the simulated environment, SE, k1 the sensor processing, SP, k2 the knowledge representation, KR while k3 the behavior generation, BG, module.

Fig. 8.7 refers to typical planning process.

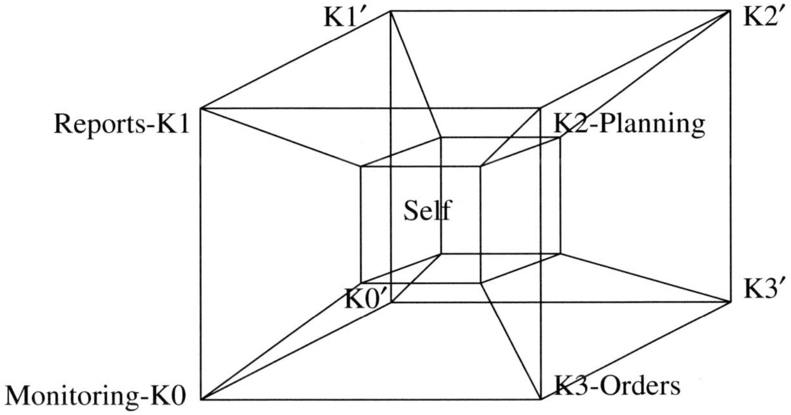

Fig. 8.7 Polytope for self-evolvable planning process

The notations for Fig. 8.7 are K0-Monitoring, K1-Reports, K2-Planning, and K3-Orders.

The Intelligent Node architecture has many applications outside the military domain, but since the work is focused on the war fighter, we therefore limit the efforts to the continuous military planning process. Military planning is a continuous process and the proposed architecture complements this process well. The architecture shown in Fig. 8.7 is designed for continuous planning triggered by incoming information.

A command, mission or orders received from a higher echelon with a clearly stated commander's intent, initiates the military decision-making process which is a planning-execution process. This process can be generalized as a sequence of steps:

A higher operational order initiates a goal-oriented collection of information via communications, generation of decisions, and contemplation of their execution.

Planning is defined as programming of the system's functioning based on the received order and the available information. As a result of planning, several courses of action are developed leading to several alternative plans. A final plan is defined as a collection of schedules for independent and/or properly distributed and synchronized processes of functioning subsystems that keep the cost functions within set boundaries. These processes are defined and distributed as sets of orders and verbal instructions to the lower echelons of the organization.

Execution starts as soon as the plan is put into action. The execution is continuously monitored.

The plan execution is monitored via the continuous interpretation of information contained in messages received from higher and lower echelons.

If analysis of this information suggests that a deviation from the original plan is taking place, and this deviation may inflict undesirable consequences, the commander and the unit-planning cell will plan for continued action.

The commander and the members of the unit planning cells are part of our Intelligent Nodes. They aid in the processes of the Decision Making or Behavior Generation elements.

The Self allows self-integrative closure and may make the entire system self-evolvable.

The alternation between direct and reverse way is beneficial for plans evaluations and creative behavior since the boundaries where creativity grows and new information is created consist of concomitant integration and differentiation tendencies.

8.4 Self-Adaptive and Self-Organizing Systems

Research into artificial self-adaptive, SA and self-organizing, SO, systems demonstrates that it is possible to develop ad hoc evolvable systems (Di Marzo Serugendo et al. 2007).

However, if we are to build self-adaptive and self-organizing, SASO, ecosystems at a large-scale or professional level, it is important to tackle issues related to their design, development and control. Indeed an open problem is how to build reliable SASO systems.

Reliability encompasses dependability properties, plus evidence of dependability. Thus, during a system's initial development, deployment and subsequent evolution, we must be able to provide assurance that emergent behaviors will respect key properties, frequently to do with safety, security or performance of the whole composed system, and that the human administrator retains control despite the self-properties of the system.

Self-adaptive, SA, systems work in a top-down manner. They evaluate their own global behavior and change it when the evaluation indicates that they are not accomplishing what they were intended to do, or when better functionality or performance is possible. Self-organizing, SO, systems work bottom-up. They are composed of a large number of components that interact locally according to simple rules.

Trustworthiness also requires acceptance by users, organizations and society at large. The challenge is imposed by the technical system global behavior of the system emerges from these local interactions, and it is difficult to deduce properties of the global system by studying only the local properties of its parts.

Self-organizing, SO, systems tend to be decentralized and bottom-up driven.

There are some important points of contact between the two concepts. The need for allowing more degrees of freedom to self-adapting systems, by allowing a level of decentralization and self-organization to the components, has already been advocated. Self-organizing, SO, systems with pure decentralized control should nonetheless provide assurance of their behavior to potential customers or users prior to deployment and should allow control to be imposed by an administrator.

Examples of systems already encompassing both self-adapting and self-organizing aspects are found in socio-technical applications involving both heterogeneous technical devices such as body or environmental sensors, software, servers, and human users such as doctors, nurses, rescue teams, end-users, and system administrators. Socio-technical systems encompass, among others, ambient intelligence and ubiquitous computing systems, emergency response or e-health applications. Each actor, human or device, in such systems is an autonomous element. As a whole the system displays complexity, self-adaptation and self-organization.

Fig. 8.8 shows the elements of a generic framework supporting engineers and developers of SASO systems.

The notations are: S-Application, K1-Metadata, K2-Reasoning, K3-Policies.

The central stage, the Self, may ensure the cooperation and redistribution of the four stages on another face of the polytope, with another starting stage.

As shown in Fig. 8.8 the reverse way should be considered too.

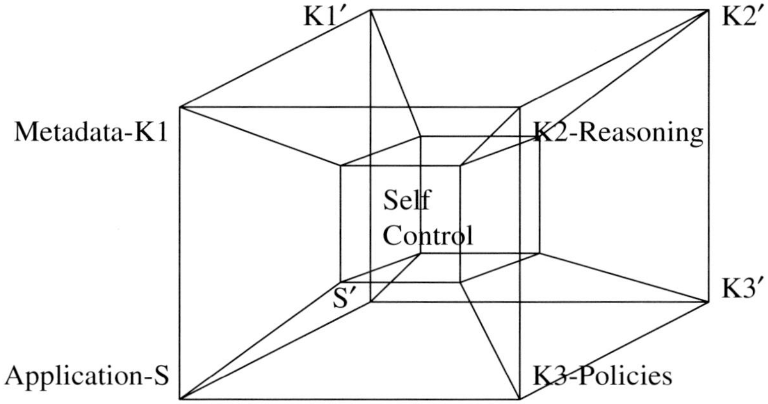

Fig. 8.8 SASO polytope

The swinging from direct to reverse way is mediated by the Self and will be beneficial for evaluation and corrections in complex problem solving.

8.5 Architecture of Evaluation System

The evaluation system represents the degree of logistic objective achievement related to the level of autonomous control. Therefore both the degree of the logistic objective achievement and the level of autonomous control must be measurable. Based on a catalog of criteria the level of autonomous control of logistics systems can be determined with adequate operations (Philipp et al. 2006, Windt et al. 2008)).

Furthermore, the logistic objective achievement can be ascertained through comparison of target and actual logistic performance figures related to the objectives low work in process, high utilization, low throughput time and high due date punctuality. The evaluation system consists of three evaluation steps to measure the logistic performance. The first step evaluates possible decision alternatives, the second step the logistic performance of individual logistic objects, as for instance orders or resources, and the third step the total system.

Evaluation step 0: Components evaluation
Evaluation step 1: Evaluation of decision alternatives
Evaluation step 2: Evaluation of individual logistic objects
Evaluation step 3: Evaluation of the total system
Fig. 8.9 illustrates an evaluation system.

Further research is directed toward the enhancement of the evaluation system to confirm the coherence between logistic objective achievement and level of autonomous control on a shop production floor as shown in Fig. 8.9. A low level of autonomous control in conventional controlled logistics systems leads to a suboptimal achievement of logistic objectives. An increase of the level of autonomous control for instance by decentralization of decision-making functions to the logistic objects, causes a rise of the achievement of logistic objectives.

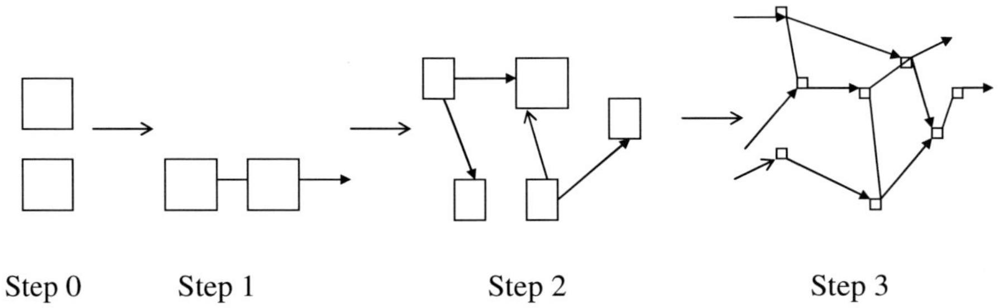

Fig. 8.9 Evaluation system

However at a certain level of autonomous control a decrease of the achievement of logistic objectives can probably be noticed caused by chaotic system behavior. By dint of simulation studies the borders of autonomous control shall be detected in order to specify in which cases an increase of autonomous control does lead to higher performance of the system. The level of autonomous control may be detected by the developed catalog of criteria which will be presented in the following chapter.

Fig. 8.10 illustrates the polytopic architecture of an evaluation system.

Arrows correspond to successive categorical levels and to successive steps shown in Fig. 8.9. The step 0 corresponds to the 0-categories. This may be associated to the objects or to areas of interest. They are called also 0-cells, or set of nodes.

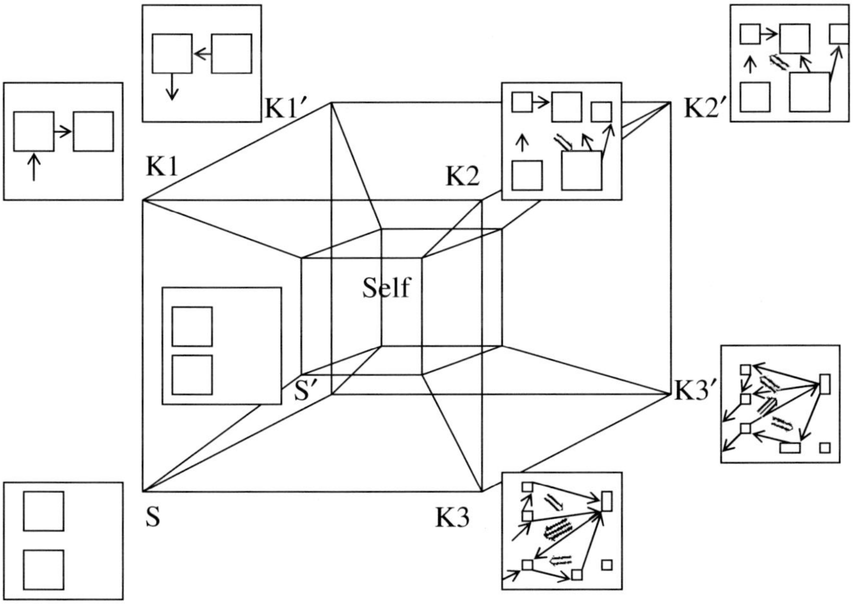

Fig. 8.10 Polytope for evaluation system

The step 1 corresponds to the 1-categories. These are illustrated by directed graphs including the morphisms that is, relations between different objects or areas of interest. The morphisms are 1-cells. They are represented here by single arrows: "$\rightarrow$". The step 2 corresponds to the 2-categories. These are illustrated by graphs plus the so-called 2-cells between paths of same source and target. The 2-cells describe relations between relations. The 2-cells are represented here by double arrows:" $\Rightarrow$". The step 3 corresponds to the 3-categories. These include

3-cells that is, the cells between 2-cells. The 3-cells are represented by triple arrows "$\Rrightarrow$". They are subjected to conditions of natural transformations.

The evaluation is considered clockwise.

Any stage embeds information the previous ones.

As shown in Fig. 8.10 the reverse evaluation way should be considered too.

The reverse way is a dual obtained by reversing the direction of arrows.

The swinging from direct to reverse developmental stages mediated by the Self will be beneficial for system self-evolvability.

Tendencies to integrate should coexist with tendencies to differentiate and it is the timing and balance of both that has a bearing.

8.6 Control Architecture and Entropy

An agent may be defined as a device or a self-directed program object which has its own value system and the means to solve certain tasks independently and then communicate its solution to a larger problem-solving organization.

The main types of agents are:

- Autonomous agents, capable of effective independent actions
- Objective directed agents, when autonomous actions are directed towards the achievement of defined tasks
- Intelligent agents, with ability to learn and adapt
- Cooperative agents, assisting other agents to perform a task

Examples of multi-agent systems are neurons in brain, antibodies in case of immune systems, ants in colonies, wolfs in packs, investors in the stock market, people in social networks, and so forth. In each case agents have relatively limited set of rules, and the complexity of the collective behavior emerges from the large number of interactions among each other and their environment. There is constant action and reaction to what other agents are doing, thus nothing in the complex system is essentially fixed.

The multi-agent architecture is significant for system functioning.

The organizations able to produce self-evolvable products would be ultimately self-evolvable ones. Modern organizations are facing complexity, chaos, interdependency, and interactions within and outside the boundaries. Global companies have begun to reorganize and decentralize their large structures, in order to enable successful strategies in spite of increasing complexity of the production problems. They have started to support organizational fluidity, with new pathways that allow project-centered groups to form rapidly and reconfigure as circumstance demand.

An open problem is how to structure the way that agents collaborate in complexity conditions. Companies are looking for a system of cooperation among groups of agents that will work significantly better than individual groups. A group contains agents working together on a common purpose. It should be partial specialization but also focus on specific tasks without losing the general objective. An important issue is how the structure and fluidity of a group will affect the global dynamics of cooperation. Fluidity depends on how easily individual agents and information can move within the company structure and how easily they can break away on their own, extending the structure. The capability to manage the complexity of modern companies depends decisively of an effective communication network.

Agility in global companies is no longer measured only by the flexibility and responsiveness of a single center but also by the agility of the network of several centers and conditioning levels.

To compare different organizations an informational entropy H calculus based on similarities as a measure of communication degree can be done. The entropy is defined by the eq. (4.4).The similarity between the groups is $(0.5)^k$ where k is the number of steps till the hierarchical level where their communication is possible. The similarity of a group with itself is 1. In this case k=0. If the groups need to move up two higher levels for communication, the similarity is 0.125. In this case k=2.

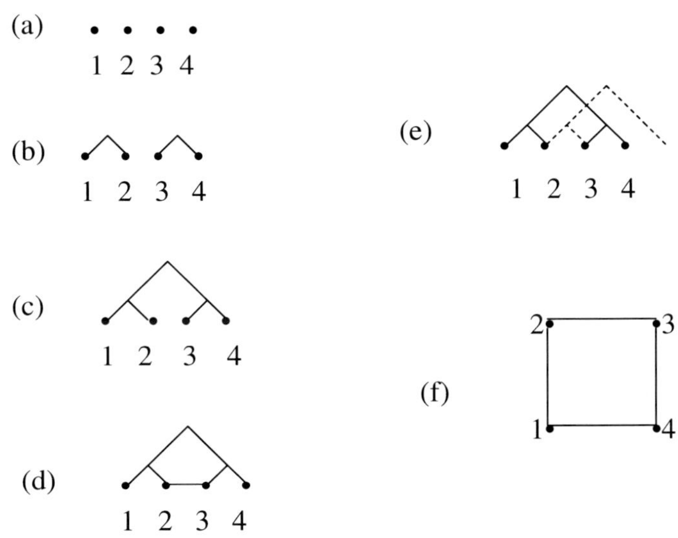

Fig. 8.11 Agents architectures

Some types of groups or individuals organizations of interest are presented in Fig. 8.11.

The illustrative examples are:

a- Isolated groups, H=0
b- Local adjacent groups, H=2.77
c- Hierarchical tree organizations, H=7.27
d- Hierarchical with horizontal associations, H=6.146
e- Multi-hierarchical groups, H=7.53
f- Multi-centric groups, H=8.31

For local organization cases the communication is inside the same group (Fig. 8.11a) or between adjacent groups (Fig. 8.11b). The proximity makes two elements to be grouped together. This does not ensure real communication between all the groups. The informational entropy is zero or very low. The entropy is H=0 for diagram shown in Fig. 8.11a, and H = 2.77 for Fig. 8.11b.

Hierarchical organizations are shown in Fig. 8.11c. The activity is performed at multiple levels and multiple scales to achieve local and global issues. In the presented example there are four groups managed by two managerial groups, directed by one directorial group. Managerial groups of different levels handle the information. The hierarchical pattern of communication is based on the following rule: the groups communicate to the managerial group that is responsible for them and the managerial groups communicate to directorial group. This allows control decisions to be made at various scales while maintaining local responsiveness and limiting the required communication. The system has low complexity but relatively high informational entropy. In this type of hierarchical architecture, a complex problem is decomposed in several simpler and smaller problems, and distributed among multiple control layers. This architecture allows the distribution of decision making among hierarchical levels. The main advantages are the robustness, the predictability and the efficiency. However the appearance of disturbances in the system reduces significantly its performances.

The modified hierarchical architectures try to find a solution to the reaction to disturbances problem, maintaining all features of hierarchical architectures and adding the interaction between modules at the same hierarchical level. This interaction allows the exchange of information between modules and improves the reaction to disturbances. Hierarchical organizations with horizontal associations are shown in Fig. 8.11d. In this case the groups establish supplementary communication patterns and the system becomes less regular and more complex. For the case presented in Fig. 8.11d one horizontal association between group 2 and 3 is included. The complexity of the system increases but the system loses entropy.

Multi-hierarchical organization is shown in Fig. 8.11e. This consists of a collection of overlapping hierarchies in which some groups pertain to more hierarchies. Groups will overlap in that some individuals can be members of several informational hierarchies. It is expected that this is better than a hierarchy in achieving stability while having in addition a position-invariant response, which allows for the control of disturbances at appropriate scale and location.

In the case of the multi-centric interactive net (Fig. 8.11f) the complete information is available to different groups making control decisions. This is not easy to do in practice for large-scale systems. These organizations allow a high performance against disturbances, but the global optimization is reduced because decision making is local and autonomous, without a global view of the system. For these reasons, a center is necessary.

A better organization would consist of groups self-organized in concentric levels around center. The information is changed easily for groups pertaining to the same level and passes through different other levels. More velocities and time scales, more sequences, for information transfer has to be considered. Teamwork may result by activity of small groups with shared purposes, diversity of skills, coherence, and communication. Complex cooperative behavior is spontaneously emergent, provided the groups are small, diverse in composition, have long outlooks and fluid structure and a pattern of interdependencies that varies. The interactive multi-centric net in continuous and recurrent reorganization movement represents a potential model for future self-evolvable organizations.

8.7 Interconnections

Computer development imposed the study of control and computation architectures (Bertsekas and Tsitsiklis 1989).

Basic communication problems in a hypercube network of processors are the problem of a single processor sending a different packet to each of the other processors, the problem of simultaneous broadcast of the same packet from every processor to all other processors, and the problem of simultaneous exchange of different packets between every pair of processors. The algorithms proposed for these problems are optimal in terms of execution time and communication resource requirements; that is, they require the minimum possible number of time steps and packet transmissions. In contrast, algorithms in the literature are optimal only within an additive or multiplicative factor.

When algorithms are executed in a network of processors, it is necessary to exchange some intermediate information between the processors. The interprocessor communication time may be substantial relative to the time needed exclusively for computations, so it is important to carry out the information exchange as efficiently as possible. There are a number of generic communication problems that arise frequently in numerical and other algorithms.

Some algorithms for the hypercube are optimal, in the sense that they execute the required communication tasks in the minimum possible number of time steps and link transmissions (Bertsekas et al. 1991).

To define a hypercube network (or d-cube), we may consider the set of points in d-dimensional space with each coordinate equal to 0 or 1. We let these points correspond to processors, and we consider a communication link for every two points differing in a single coordinate. Thus we obtain an undirected graph with the processors as nodes and the communication links as arcs. The binary string of length d that corresponds to the coordinates of a node of the d-cube is referred to as the identity number of the node.

A hypercube of any dimension can be constructed by connecting lower dimensional cubes, starting with an l-cube. In particular, we can start with two (d -1)-dimensional cubes and introduce a link connecting each pair of nodes with the same identity number. This constructs a d-cube with the identity number of each node obtained by adding a leading 0 or a leading 1 to its previous identity, depending on whether the node belongs to the first (d -1)-dimensional cube or the second (see Fig. 8.12). We may refer to a d-cube node interchangeably in terms of its identity number, a binary string of length d, and in terms of the decimal representation of its identity number. Thus, for example, the nodes (00…0), (00…1) and (11…1) are also referred to as nodes 0, 1, and 2^d-1, respectively.

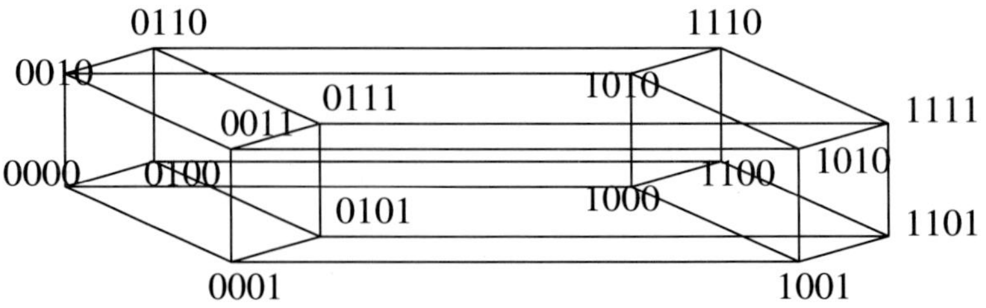

Fig. 8.12 Construction of the 4-cube

Fig. 8.12 shows the construction of 4-cube.

The Hamming distance between two nodes is the number of bits in which their identity numbers differ. Two nodes are directly connected with a communication link if and only if their Hamming distance is unity, that is, if and only if their identity numbers differ in exactly one bit. The number of links on any path connecting two nodes cannot be less than the Hamming distance of the nodes. Furthermore, there is a path with a number of links that is equal to the Hamming distance, obtained, for example, by swinging in sequence the bits in which the identity numbers of the two nodes differ or equivalently, by traversing the corresponding links of the hypercube. Such a path is referred to as a shortest path in this paper and a tree consisting of shortest paths from some node to all other nodes is referred to as a shortest path tree.

Information is transmitted along the hypercube links in groups of bits called packets. For such algorithms it is assumed that the time required to cross any link is the same for all packets and is taken to be one unit. Thus, the analysis applies to communication problems where all packets have roughly equal length. Packets can be simultaneously transmitted along a link in both directions and that their transmission is error free.

Only one packet can travel along a link in each direction at anyone time; thus, if more than one packet is available at a node and is scheduled to be transmitted on the same incident link of the node, then only one of these packets can be transmitted at the next time period, while the remaining packets must be stored at the node while waiting in queue.

There are several kinds of regular graphs described in the literature that find applications for interconnections (Oh and Chen 2001, Patil et al. 2009). In a connected m-regular graph, each node has degree m. Regular graphs are interesting to us because of their symmetric load distribution. In such experiments, when load distribution is given high importance, topologies tend toward regular graphs. A hypercube graph is a regular graph of 2^m nodes, represented by all m-length binary strings. Each node connects to all other nodes that are at a Hamming distance of 1, forming an m-regular graph. A hypercube graph has a diameter of m, which is the maximum Hamming distance between any two nodes.

The De Bruijn graph is a directed graph where each node is mapped onto an identifier in the identifier space formed by all m-length strings of an alphabet of length b. Every node has exactly m outgoing edges.

The m edges are drawn by right shifting each node identifier by one position, and adding each of the b symbols in the alphabet at the end.

A detailed graph-theoretic analysis of peer-to-peer networks, with respect to routing distances and resilience to faults, was provided (Loguinov et al. 2005). The paper argues that De Bruijn graphs offer the optimal diameter topology among the class of practically useful graphs because of their low diameter.

The HyperCup shown in Fig. 8.13 is a hypercube graph constructed in a distributed manner by assuming that each node in an evolving hypercube takes more than one position in the hypercube (Schlosser et al. 2002). That is, the topology of the next dimensional hypercube implicitly appears in the present hypercube, with some of the nodes also acting as virtual nodes to complete the hypercube graph.

Similarly, when nodes go away, some of the existing nodes take the missing positions along with their own.

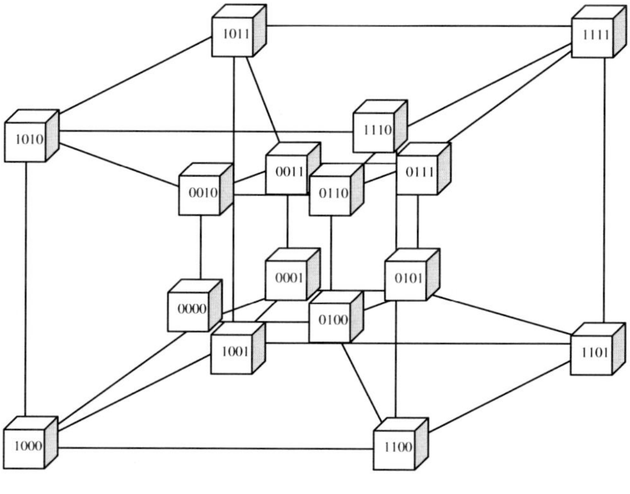

Fig. 8.13 HyperCup

References

Arkin, R.: Behaviour-based robotics. Bradford Book, MIT Press (1998)

Bertsekas, D.P., Tsitsiklis, J.N.: Parallel and Distributed Computation: Numerical Methods. Prentice-Hall, Englewood Cliffs (1989)

Bertsekas, D.P., Ozveren, C., Stamoulis, G.D., Tseng, P., Tsitsiklis, J.N.: Optimal Communication Algorithms for Hypercubes. J. Parallel Distrib. Comput. 11, 263–275 (1991)

Dawidowicz, E.: Intelligent Agent Technology in Command and Control Environment. In: Proceedings of 1999 Command and Control Research and Technology Symposium, Newport, Rhode Island, June 29-July 1 (1999)

Dawidowicz, E.: Performance Evaluation of Network Centric Warfare Oriented Intelligent Systems. In: Proceedings of PERMIS 2001, Gaithersburg, MD (2002)

Di Marzo Serugendo, G., Fitzgerald, J., Romanovsky, A., Guelfi, N.: A Generic Framework for the Engineering of Self-Adaptive and Self-Organising Systems. Technical Report CS-TR-1018. School of Computing Science. University of Newcastle (2007)

Loguinov, D., Casas, J., Wang, X.: Graph-Theoretic Analysis of Structured Peer-to-Peer Systems: Routing Distances and Fault Resilience. IEEE/ACM Transactions on Networking 13(5), 1107–1120 (2005)

Luzeaux, D., Dalgalarrondo, A., Dufourd. D.: Autonomous small robots for military applications. In: Proceedings of UVS Tech 2001, Bruxelles, Belgique (2001)

Maturana, H., Varela, F.: The tree of knowledge: The biological roots of human understanding. Shambala, Boston (1992)

Oh, E., Chen, J.: Parallel routing in hypercube net-works with faulty nodes. In: IEEE International Conference on Parallel and Distributed Systems (ICPADS 2001), pp. 338–345 (July 2001)

Patil, S., Srinivasa, S., Mukherjee, S., Rachakonda, A.R., Venkatasubramanian, V.: Breeding Diameter-Optimal Topologies for Distributed Indexes. Complex Systems 18(2), 175–194 (2009)

Philipp, T., Böse, F., Windt, K.: Evaluation of autonomously controlled logistic processes. In: Proceedings of 5th CIRP International Seminar on Intelligent Computation in Manufacturing Engineering, pp. 347–352 (2006)

Schlosser, M.T., Sintek, M., Decker, S., Nejdl, W.: HyperCuP - Hypercubes, Ontologies, and Efficient Search on Peer-to-Peer Networks. In: Moro, G., Koubarakis, M. (eds.) AP2PC 2002. LNCS (LNAI), vol. 2530, pp. 112–124. Springer, Heidelberg (2003)

Sanz, R., Lopez, I., Bermejo-Alonso, J., Chinchilla, R., Conde, R.: Self-X: The control within. In: Proceedings of IFAC World Congress 2005 (2005)

Sanz, R., López, I., Rodríguez, M., Hernández, C.: Principles for consciousness in integrated cognitive control. Neural Networks 20(9), 938–946 (2007)

Sanz, R., Hernández, C., Sánchez, G.: Consciousness, meaning and the future phenomenology. In: Machine Consciousness 2011: Self, Integration and Explanation - AISB, York, UK (2011)

Windt, K., Philipp, T., Böse, F.: Complexity Cube for the Characterization of Complex Production Systems. Int. J. Comput. Integr. Manuf. 21, 195–200 (2008)

Chapter 9
Manufacturing Systems

Abstract. The development of manufacturing systems from fixed to flexible, reconfigurable and self-evolvable systems with reference to assembly operations is outlined.

The perspectives of polytopic models for self-manufacturing are evaluated.

Informational entropy criteria are used to evaluate manufacturing strategies and to characterize supply chain networks.

9.1 Viable Systems Models

The viability of complex systems through processes of self-regulation, self-organization and control was studied by Beer (Beer 1985).

Knowledge cybernetics is a related approach principally concerned with the development of agents like autonomous social collectives that survive through knowledge and knowledge processes (Yolles 2006).

Fig. 9.1 shows an example of viable system architecture.

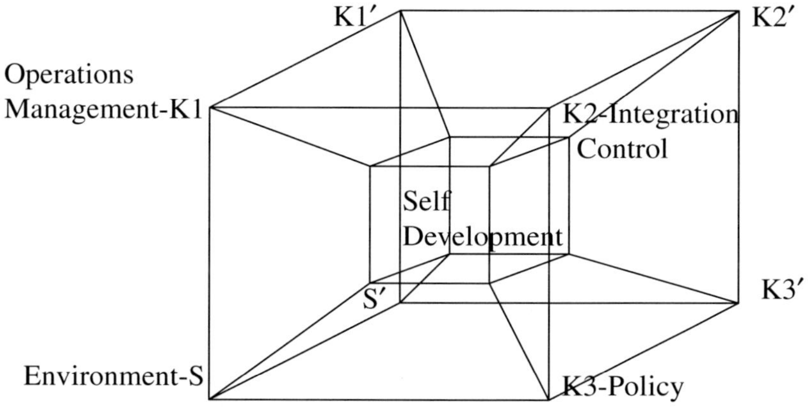

Fig. 9.1 Polytope for viable systems

For the architecture shown in Fig. 9.1, the levels may be identified as: S-Environment, K1-Management, K2-Control and Coordination and K3-Policy. The Development module, linking the previous levels, is critical for viability.

The Self is considered as the starting area of one life cycle of the system.

The frame of viable systems modeling architecture has demonstrated a great deal of potential in creating ways of analyzing complex situations and demonstrated a possibility to be used to diagnose complex situations and to be used as a means by which improvement can be engineered.

The fractal structure of the viable system model as outlined by Fig. 9.1 means that the same mechanisms are replicated at level and in each of the sub-systems and sub-sub-systems in the unfolding of complexity.

This means for instance, that decision-making should be a multi-level activity, and not merely the prerogative of management as in a command and control system. This allows for strategy to be built up through the organization as a series of conversational processes between different levels, so that the strategy for the organization as a whole both informs and is informed by the planning at divisional level. Similarly, divisional strategy both informs and is informed by departmental strategy, and so on down to the level of teams and individuals.

There exist a construction way S→K1→K2→K3 and a deconstruction way K3′→K2′→K1′→S′ that should be considered from early designs. The reconciliation of these two ways is a negotiated process since weaknesses in planning should be avoided. The boundaries where innovative solutions appear consist of synchronized tendencies. Tendencies to construct should coexist with tendencies to deconstruct and it is the timing and the balance of both that would make a difference.

The two ways are considered as significant for long-term strategic alliances (Camarinha-Matos and Afsarmanesh 2005). The swinging between construction and deconstruction will ensure viability for unexpected market fluctuation or critical situations.

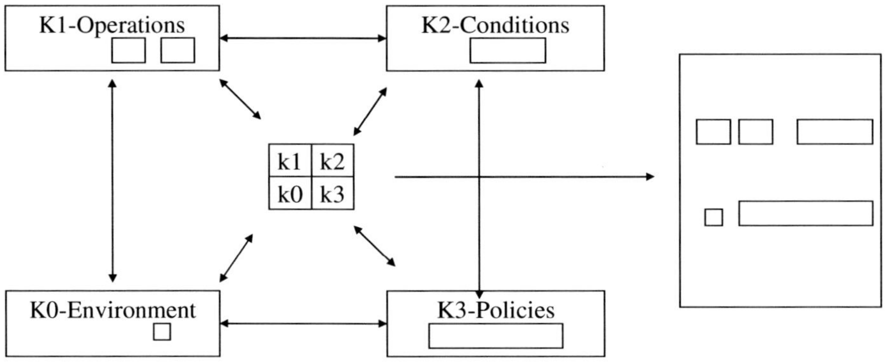

Fig. 9.2 Operad for viable systems

Fig. 9.2 shows the operad associated to viable system architecture.

The notations are: K0-Environment, K1-Operations, K2-Conditions, and K3-Policies.

In this case k0, k1, k2 and k3 activate K0, K1, K2 and K3 respectively.

This takes into account the fact that in viable systems each level has a different perspective, focus of attention and often a different time horizon.

9.2 Self-Evolvable Manufacturing Systems

Recent road mapping efforts have all clearly underlined that true industrial sustainability will require far higher levels of systems' autonomy and adaptability. In accordance with such recommendations, the Self-Evolvable Production Systems, SEPS, has aimed at developing such technological solutions and support mechanisms. Since its inception as a next generation of production systems, the concept is being further developed and tested to emerge as a new production system paradigm. The essence of self-evolvability resides not only in the ability of system components to adapt to the changing conditions of operation, but also to assist in the evolution of these components in time such that processes may have self-properties.

Typically, self-evolvable systems have distributed control and are composed of intelligent modules integrated. To assist the development and life cycle issues, comprehensive methodological framework is being developed. A concerted effort is being exerted through research projects in collaboration with manufacturers, technology/equipment suppliers, and research institutions.

The globalization of markets, shortening of product life cycles, decrease of dimensions for products and outsourcing were identified as major threats for industry. Answers to such threats were paradigms as evolvable assembly system, EAS (Onori 2002), evolvable production systems, EPS (Onori et al. 2006, Frei et al. 2007), and evolvable manufacturing systems, EMS (Iordache 2009).

The design process of assembly systems, EAS, has been modeled by a hierarchy of four levels: S-Environment, K1-Domain knowledge, K2-Inference knowledge and K3-Task knowledge (Lohse et al. 2005). The domain knowledge level defines all the specific concepts needed by the inferences. The inference knowledge level defines what inferences are needed to fulfill the reasoning task. The task knowledge level defines the reasoning tasks required to achieve a specific goal.

The four-levels hierarchy does not allow complete evolvability and autonomy.

EPS represents a concrete solution to the requirements from the market such as stated within the agile, reconfigurable and distributed approaches. They include high responsiveness, low down-times, ability to handle small series with many variants, and changeability. Together with ontology-based process specific modules, a distributed control system using the multiple agent paradigm allows to quickly and cost effectively adapt to ever-changing production requirements.

EPS have similarities with the bionic, fractal, holonic, biological and reconfigurable manufacturing systems, but there exist major differences too.

Besides considering system morphology, EPS strongly links product, processes, system and environment by means of detailed ontologies.

EPS focuses on high-level organization and implies the ability of complex systems to co-evolve with continuously changing requirements. EPS are expected to allow the user to build any required system and to modify this at wish.

Some features of the production systems necessary to achieve evolvability are:

- Modularity since many small, dedicated units that can easily be integrated into different systems/cells
- Process orientation for units
- Multiple scales, variable granularity and fluidity process related
- Distributive control system
- Interoperability
- Use of the multi-agent technology to capture emergent behavior

Table 9.1 compares the conventional systems with the self-organized multi-agent systems.

Table 9.2 shows categorification aspects for management systems.

Self-evolvable systems may be considered as a natural development of flexible, reconfigurable and evolvable manufacturing systems.

Table 9.1 Self-organized multi-agent versus conventional systems

Characteristics	Conventional	Self-organized Multi-agent systems
Model source	Military	Biology, Sociology
Optimum	Yes	No
Prediction level	Individual	Self-Aggregate
Computational stability	High	Low
Match to reality	Low	High
Requires central data	Yes	No
Response to change	Fragile	Relatively robust
System reconfiguration	Hard	Easy
Calculus	Complicated, long	Simple, short
Time to schedule	Slow	Real time
Processing	Sequential	Concurrent, parallel

Table 9.2 Categorification for management systems

Criterion\ System	Specialized	Flexible	Re-configurable	Evolvable	Self-Evolvable
Level	K0	K1	K2	K3	Self
	n=0	n=1	n=2	n=3	n≥4
Categories	0-category	1-category	2-category	3-category	4-category
Skills	One	Set of fixed skills	More skills adapted	No particular product focus	No particular product focus
Flexibility	Low	Discrete	Continuous	Emergent	Multiple Emergent
Capability	High efficiency	Cope with different situations	Cope with differences. Adaptable	Agile	Agile Anticipative Opportunist
Concerns	Rigid	Cannot cope with new	Unexpected are not coped	Difficult to define generic mechanism	Self-develops generic mechanisms

Table 9.2 suggests considering the different stage in the historical development of manufacturing systems as the necessary stages in categorification.

The first stage corresponds to specialized manufacturing, to single installation and in the same time to sets or 0-categories.

A first-order evolvability step is represented by the transition to flexible manufacturing systems.

Flexibility approach allows doing diverse tasks with the same installation. This may be linked to 1-categories.

A second-order evolvability step is represented by the transition to reconfigurable manufacturing systems.

Reconfiguration is supposed to make use of several installations. It is linked to the 2-categories. Reconfigurable manufacturing systems incorporate principles of modularity, integrability, flexibility, scalability, convertibility and diagnosability. Some flexible and reconfigurable systems failed because they do not take into account that if any system is to be flexible then its constituents need to be far more flexible.

A third-order evolvability step is represented by the transition to evolvable manufacturing systems. Evolvability achieves the full flexibility and is related to the 3-categories concept implementation.

Interrelated with self-evolvability is the notion of agility.

Agility is the ability of an organization to adapt to change and also to seize opportunities that become available due to change.

The problem of how organizations can successfully deal with unpredictable, dynamic, and constantly changing environments has been a prevailing topic both in industry and academia for a few decades. Many different solutions have been proposed: networking, reengineering, modular organizations, virtual corporations, high performing organizations, employee empowerment, flexible manufacturing, just-in-time, and so on. Among proposals of how to deal with an uncertain and unpredictable environment, the three notions of *adaptive organization*, *flexible organization* and *agile enterprise* are the most predominant and popular. There are many different approaches to define each of these terms and there is much confusion and ambiguity concerning definitions and components of each of these concepts. Some authors make a sharp differentiation between those concepts while others use them synonymously. However, in general all concepts were considered as possessing the ability to adjust and respond to change.

Observe that EAS, EPS, EMS considers the production unit as an artificially living entity and emphasizes on evolution rather than adaptation.

Usually the adaptability implies an adjustment on the time scale of the life cycle of the organism. It characterizes 1-category frames. But this is not enough to challenge the high complexity. Evolvability should imply the capacity for genetic-like change to invade new life-like cycles on several time scales, by n-categorification steps.

In a dynamic environment, the lineage that adapts first wins. Fewer mutation steps mean faster evolution. The request is for some production or management systems built to minimize the number of mutations required to find improvements.

By successive categorification steps the legacy equipment and associated software will still be utilizable.

Categorical issue implies that EMS achieves specific fluidity properties. It should have fluidity at different levels of complexity. Consider that the production line is composed of several components that can be plugged in or out. These are 1-cells and the corresponding fluidity is the so-called fine fluidity or 1-fluidity corresponding to flexible manufacturing and to 1-categories.

When a manufacturing line is composed of several cells and these cells are modules or 2-cells that can be plugged in or out this is the thin fluidity called also 2-fluidity. It corresponds to reconfigurable manufacturing and to 2-categories.

The thick fluidity or 3-fluidity will refer to the whole system that is 3-cells to be plugged in or out. This corresponds to evolvable manufacturing and to 3-categories.

The autonomic and the organic computing have been identified as fundamental concepts for achieving evolvable manufacturing systems. Although autonomic computing was designed for software systems, the related ideas can be projected into a modular production system. Automatic computing in this context refers to computing elements disseminated throughout the production system which beyond the normal mechanical, electrical and sensorial units includes computational power.

Organic computing focuses on completing the closure by studying a 4th order evolvability step.

The study of management systems for self-integrative closure and the emergence of the Self as self-management systems corresponding to n≥4 represent a challenge.

Fig. 9.3 outlines a four sub-realms network for evolvable manufacturing systems

The notations are: K0-Environment, K1-Products, K2-Systems, and K3-Processes.

The environment refers to real and artificial aspects, including the available materials.

Products sub-realm denotes the products and product-related activities. Production sub-realm denotes the production system skills, modules. Processes sub-realm refers to all processes, for example assembly.

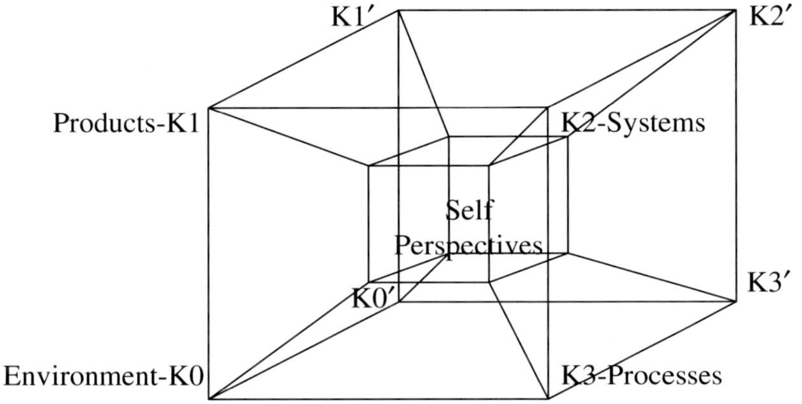

Fig. 9.3 Polytope for self-evolvable manufacturing systems

Onori highlighted the interaction between products and systems illustrated by a generic product life cycle view (Onori 2002).

Observe that the construction of a specific self-evolvable manufacture parallels and recapitulates the general history of manufacturing systems from specialized to self-evolvable.

Fig. 9.4 describes the Adacor architecture of the manufacturing polytope (Leitao and Restivo, 2006). The notations are: K0-Physical, K1-Operational, K2-Management, K3-Planning.

ADACOR architecture is built upon a set of autonomous and cooperative holons, to support the distribution of skills and knowledge, and to improve the capability of adaptation to environment changes. Each holon is a representation of a manufacturing component that can be either a physical resource as numerical control machines, robots, programmable controllers, and pallets or logic entity products, orders, and so on.

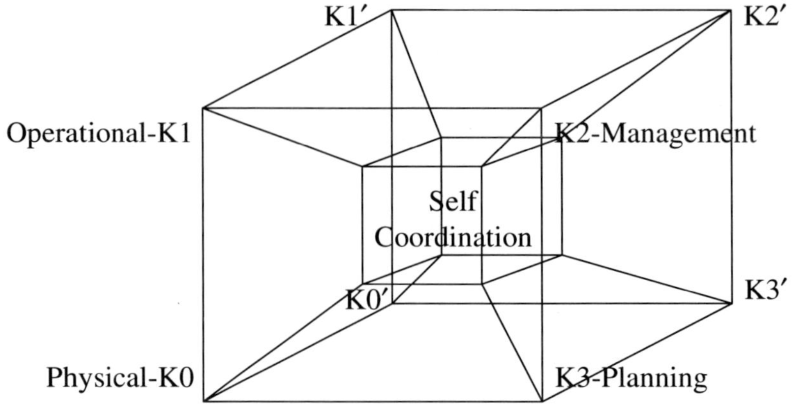

Fig. 9.4 Polytope for Adacor architecture

ADACOR holons perceive their environment and response quickly to changes, reacting to the stimulus provided by the environment. In spite of their predominant reactive behavior, ADACOR holons do not simply act in response to their environment, but they are also able to take the initiative, for example elaborating product plans or predicting the occurrence of future disturbances. Such holons are of the plug and produce type, being possible to add a new element without the need to re-initialize the system, thus allowing high flexibility in system adaptation and re-configuration.

ADACOR architecture defines four manufacturing holon classes, product (K3), task (K2), operational (K1) and supervisor holon Self classes, according to their functions and objectives. The supervisor holon introduces coordination and global optimization in decentralized control and is responsible for the formation and coordination of groups of holons.

Each product available to be produced in the factory plant is represented by a product holon, containing all information related to the product and being responsible for the short-term process planning. The product holon acts as the bridge between the shop floor and planning levels, contributing to the integration of all the manufacturing control functions, that is planning, scheduling and plan execution.

As shown in Fig. 9.3 and Fig. 9.4 the reverse ways should be considered too.

This means that after the integrative way S→K1→K2→K3 we need to look at the differentiation way K3′→K2′→K1′→S′.

Making use of the developments of the direct way, new information will result by reverse way. The swinging from direct to reverse developmental stages mediated by the Self is beneficial for evaluation and corrections of manufacturing system since the boundaries where new solutions are created consist of simultaneous tendencies. Tendencies to integrate coexist with tendencies to differentiate and it is the delicate mix of both that may be significant for self-evolvability.

9.3 Self-Reconfigurable Systems Structure

In order to make manufacturing systems more adaptive, the Reconfigurable Manufacturing Systems (RMS) concept was introduced (Hu and Efstathiou 2007).

In a reconfigurable system, hardware and software are divided into modules, with a swinging control architecture, which means functionality and capacity can be adjusted as appropriate.

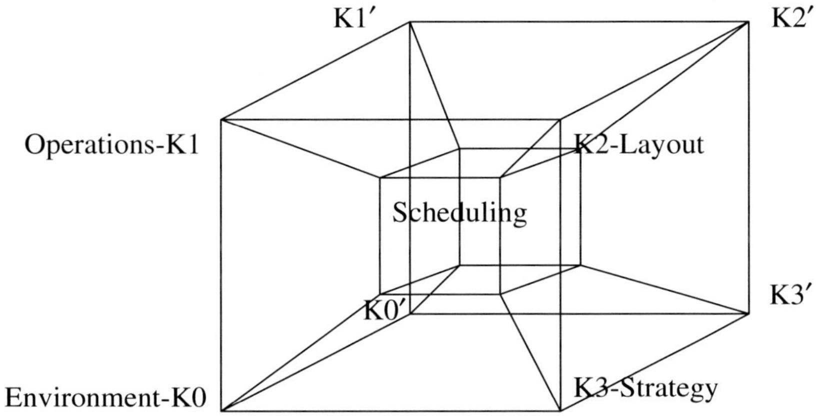

Fig. 9.5 Polytope for reconfigurable systems

Hu and Efstathiou propose using the inherent modularity of RMS to construct the three-level architecture, as shown in Fig. 9. 5.

The notations are: K0-Environment, K1-Operations, K2-Layout, and K3-Strategy.

The three levels are organized in a way that makes inter-level communication direct and efficient.

On level K1, the operations level, a Demand Analyzer is introduced. It receives customer orders and decides whether to trigger Rescheduling in order to handle disruptions. The Scheduling module will decide how to distribute the tasks among machine tools. Then each task is processed in Work In Progress, WIP module and shipped to market.

On level K2, the layout levels, as the Reconfiguration Module and Product Family Management, PFM, may be introduced based on entropy measurement. Thus, we can adjust functionality and capacity according to demand trends. PFM bases product selection decisions on demand patterns, and decides which products and services to provide.

On level K3, the strategy level, the Trend Detector analyzes market demand and reports any noticeable pattern. Information is then passed to level K2 to arrange for reconfiguration once necessary.

Enterprises can also use pricing policy to affect demand as well as customer behavior.

Overall, level K1 and level K2 correlate to each other the same way as that of adaptive control – two different feedbacks at different pace, both event-triggered. Level K3 supervises them and uses PFM and forecasting to make reconfiguration arrangements. This would enable dynamic scheduling as well as product transition in the job shop at relatively low cost. The system can be adaptive to customer needs and market fluctuations.

The feasibility and justification of implementing reconfigurable manufacturing system by redefining production objectives is a critical problem.

This means to look at the configuration way K0→K1→K2→K3 and the reconfiguration way K3′→K2′→K1′→K0′.

The Self module corresponding to scheduling should modulate the interaction of the direct and reverse ways. To give an example, the two-way interaction suggests designing, building and maintaining automated plants to avoid off-shoring tasks or entire industries and making durable investments in a way that will not be wasted when the product market or context evolves.

Tendencies to configuration should coexist with tendencies to reconfiguration and it is the rhythm and balance of both that is relevant for self-evolvability.

Another self-reconfigurable manufacturing system was studied by Mun (Mun et al. 2004).

It is based on fractal manufacturing systems, FrMS.

To respond quickly to the rapidly changing manufacturing environment, it is imperative for the system to have such capabilities as flexibility, adaptability, and reusability. The fractal manufacturing system, FrMS, is a new manufacturing paradigm designed to meet these requirements. Through its dynamic restructuring process the FrMS is able to reorganize the system's configuration by its definition. To facilitate a dynamic reconfiguration of system elements, the fractals as well as software modules should be self-reconfigurable. Embodiment of a self-reconfigurable manufacturing system is achieved by using self-reconfigurable software architecture.

Self-reconfigurable software architecture is designed by conducting the following studies:

- Analysis of functional requirements of a fractal and environmental constraints.
- Design of reconfigurable software architecture.
- Selection of proper techniques to implement software modules, and realization of software architecture equipped with self-reconfiguration capability.

To validate this approach, the designed architecture is applied to the FrMS.

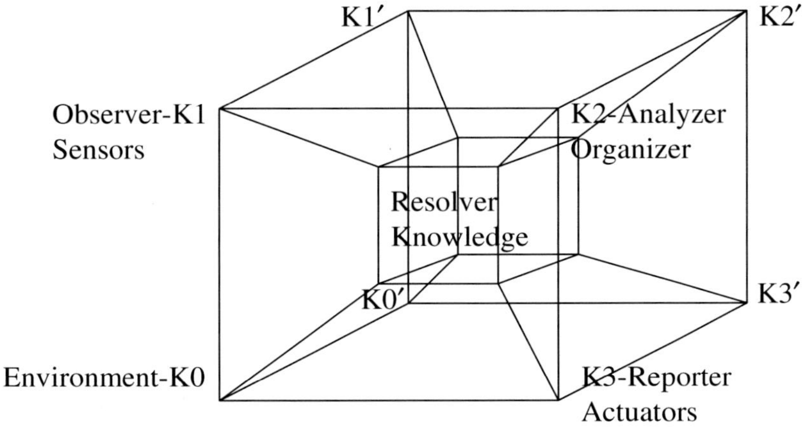

Fig. 9.6 Polytope for fractal architecture

A fractal architecture consisting of several functional modules as observer, analyzer, resolver, organizer and reporter and several auxiliary modules for helping the main modules was studied by Ryu (Ryu and Jung 2003). It is specifically designed to represent the elements at any level in the system hierarchy and its cooperation or interaction with adjacent levels. Fractals at any level have the same functional modules to support their operations and achieve their goals. Among the five modules, the observer and the reporter function as a gate to communicate with other fractals. By constituting the five function modules and facilitating coordination, cooperation and negotiation among those modules, the characteristics of a fractal can be applied into the system to achieve the shop-level goal. Fig. 9.6 shows the architecture of the bottom-level fractal and relationships among functional modules.

In this case the notations are: K0-Environment, K1-Observer Sensors, K2-Analyzer Organizer, and K3-Reporter Actuators.

The Self module consists of resolver and knowledge database. These appear in Fig. 9.6 as instances of the Self.

The fractality of the system means that architecture of the Self module parallels that of the external cube in Fig. 9.5.

9.4 NK-Model for Manufacturing

The NK-model can be used to better understand strategy formulation as complex adapting system of capabilities and to recognize the interaction between capabilities and competing strategies.

The system to be studied is a manufacturing strategy (McCarthy 2004). It is analyzed and coded as a string of elements, N, where each element is a capability. For any element i, there exist a number of possible states which can be coded using integers 0, 1, 2, 3, and so on. The total number of states for a capability is described as A_i. Each system, or strategy, s is described by the chosen states

$s_1s_2...s_N$ and is part of an N-dimensional landscape or design space, S. The K parameter in the NK-model indicates the degree of connectivity between the system elements or capabilities. It suggests that the presence of one capability may have an influence on one or more of the other capabilities in a company manufacturing strategy.

To understand the significance of this design space to manufacturing strategy formulation, a seminal example is adopted and conceptually modified from Kauffman's work (Kauffman 1993, McCarthy 2003).

Table 9.3 highlights NK-model potentialities.

Table 9.3 shows the NK-model notation and outlines its relevance to manufacturing strategy if compared to evolutionary biology.

Table 9.3 NK-model potentialities

Notations	Biology	Manufacturing
N	Number of genes	Parts, components
K	Epistatic interactions	Interconnectedness
A	Alleles that a gene	Possible states, properties
C	Coupled genotypes	Co-evolvability

The design space is A^N, which provides the number of possible manufacturing strategies, each of which is allocated a random fitness between 0 and 1.

A value close to 0 indicates poor fitness, while a value close to 1 indicates good fitness. In principle, the fitness values can then be plotted as heights on a multidimensional landscape, where the peaks represent high fitness and the valleys represent low fitness.

In Kauffman's model, the fitness function F (x) is the average of the fitness contributions, f_i (x), from each element i, and is written as:

$$F(x) = \frac{1}{N}\sum_{i=1}^{N} f_i(x) \tag{9.1}$$

At N=4 a four-dimensional frame cube can be used to represent the possible combinations and their relationships to each other.

Fig. 9.7 refers to strategic options generated by four capabilities: cost, quality, flexibility and delivery.

Fig. 9.7 shows the polytopic model for co-evolutionary manufacturing.

It uses a binary notation to represent the presence (1) or absence (0) of a capability. For example, strategy 0011 indicates that the capabilities flexibility and delivery are present, while the capabilities cost and quality are absent. The base strategy 0000 is at the top of the diagram, while the maximum strategy 1111 is at the bottom of the diagram.

As a manufacturing company strategy aggregates additional capabilities, it descends into the lower parts of the diagram. Lines are used to connect two immediate neighbors.

Each corner point of the 4-cube represents a manufacturing strategy and its hypothetical fitness value. Strategic change is assumed to be a process of moving from one strategy to another in search of an improved fitness. This is known as the adaptive walk. If we arbitrarily select a point on the cube, for instance point 1011, there are three one-mutation neighbors. These are points 1010, 1111 and 1100. If point 1010 has an immediate neighbor strategy with a higher fitness value then it is possible that a manufacturing firm would evolve to this strategy. A local peak is a strategy from which there is no fitter point to move to in the immediate neighborhood. A global peak is the fittest strategy on the entire landscape.

Let us consider the high capabilities rectangle in Fig. 9.6 has a digit representation: [1111], [1110], [1011] and [1010].

The set of four vectors corresponds to a 4x4 matrix and based on this we can use entropic analysis.

To this matrix one may associate a similarity matrix. The entropy associated to this matrix is H (High) = 6.475.

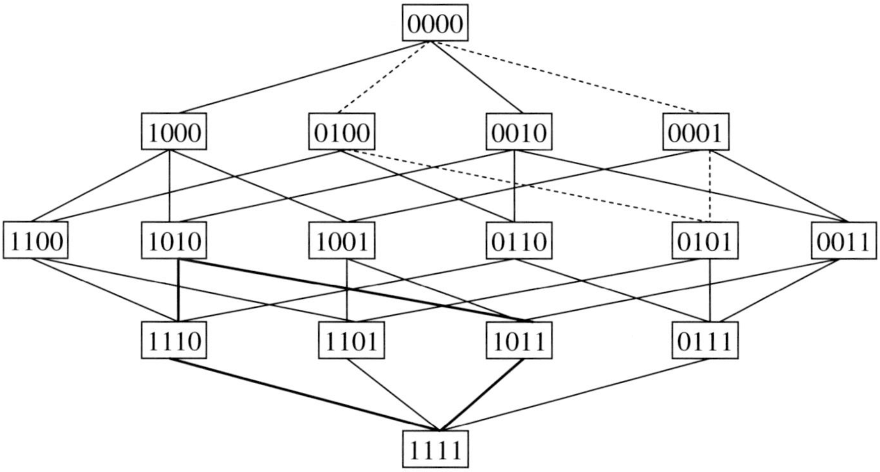

Fig. 9.7 Four manufacturing capabilities polytope

By adding new vectors to this reference state of 4 vectors the entropy H varies.

There is only a small change of entropy, ΔH if the vector of the test is similar to the reference set and this supplementary option is though to have similar properties.

If a database shares similar bit patterns with reference set, adding a similar vector will induce a change targeting the minimum entropy production.

By contrast, inclusion of an option having dissimilar vector leads to a higher entropy production, targeting the maximum entropy production.

In this way the options may be screened to identify strategies that cause low or high changes of the reference set informational entropy and detect other promising options according to the established goal.

The tested directions are [0011], [0111], [1001], [1101], [1000], [1100], [0010] and [0110].

Denote by High the matrix of strong capabilities.

$$\text{High} = \begin{vmatrix} 1 & 1 & 1 & 1 \\ 1 & 1 & 1 & 0 \\ 1 & 0 & 1 & 1 \\ 1 & 0 & 1 & 0 \end{vmatrix} \tag{9.2}$$

The results are shown in Table 9.4.

Table 9.4 shows the informational entropy for high capabilities

Table 9.4 Informational entropy for high capabilities

New Step Matrix	High	High [0011]	High [0111]	High [1101]	High [1101]	High [1000]	High [1100]	High [0010]	High [0110]
H	6.475	11.958	11.958	10.894	10.894	10.894	10.894	11.958	11.958
ΔH	0	5.483	5.483	4.419	4.419	4.419	4.419	5.483	5.483
Δ^2H	-	0	-1.064	0	0	0	1.064	0	-

The low-level rectangle shown in Fig. 9.6 has the digit representation: [0000], [0001], [0100] and [0101].

Denote by Low the matrix of low capabilities.

$$\text{Low} = \begin{vmatrix} 0 & 0 & 0 & 0 \\ 0 & 0 & 0 & 1 \\ 0 & 1 & 0 & 0 \\ 0 & 1 & 0 & 1 \end{vmatrix} \tag{9.3}$$

Table 9.5 shows the informational entropy for low capabilities

Table 9.5 Informational entropy for low capabilities

New step matrix	Low	Low [0011]	Low [0111]	Low [1001]	Low [1101]	Low [1000]	Low [1100]	Low [0010]	Low [0110]
H	6.475	10.894	10.894	11.958	11.958	11.958	11.958	10.894	10.894
ΔH	0	4.419	4.419	5.483	5.483	5.483	5.483	4.419	4.419
Δ^2H	-	0	1.064	0	0	0	-1.064	0	-

The periodicity for $\Delta^2 H$ is observed. This should be correlated to market periodicities.

9.5 Entropy for Supply Chain Networks

A supply network consists of several organizations, being linked by materials, information and financial flows. These organizations may be firms producing parts, components and end products, logistic service providers or even customers. The supply chain management is deputy manager to the coordination of all these distinct processes in the most efficient way. Supply chain management means transforming a company's supply chain into an optimally efficient, customer-satisfying process, where the effectiveness of the whole supply chain is more important than the effectiveness of any individual department. A supply chain, however, is not a simple linear sequence of connections, but rather, an intricate web-like structure. We are therefore operating with a complex network of relations and connections between different partners. On the one hand, market globalization and products variety were required by customers.

Manufacturing systems are reorganizing and re-engineering themselves to respond to the rapidly changing market. Systems are required to modify frequently to meet the uncertain demand. Modification could be adding or removing processes, reallocation of working resources and changes to buffer sizes.

Complexity reduction as a strategic goal for the operation has been investigated and measured by previous works in this field.

Literature dealing with this topic includes analytical approaches to measure the complexity of supply chains and manufacturing systems (Karp and Ronen 1992; Frizelle and Suhov 2001).

The main approaches towards measuring system complexity are based on entropy measures: information-theoretic modeling of manufacturing organizations has led to the development of an entropic method to compute the static and dynamic complexity measure of a single manufacturing system (Gino 2002, Battini et al. 2007, Wang et. al 2005).

A production network of cells as that shown in Fig. 9.8 may be evaluated by entropy calculus.

The transfer coefficients from compartment i to compartment j, denoted by f_{ij}, are of the order of units (f_{41}, f_{56}, f_{57}, f_{67}), of the order of 10^{-1} (f_{18}, f_{43}), and of the order of 10^{-2} (f_{21}, f_{31}, f_{23}, f_{85}, f_{89}). Other coefficients being smaller than 10^{-2} are considered as nulls.

The similarities r_{ij} will be defined by $r_{ij}= 2^m$ where m=0, -1, -2 is the order of the transfer coefficient f_{ij}.

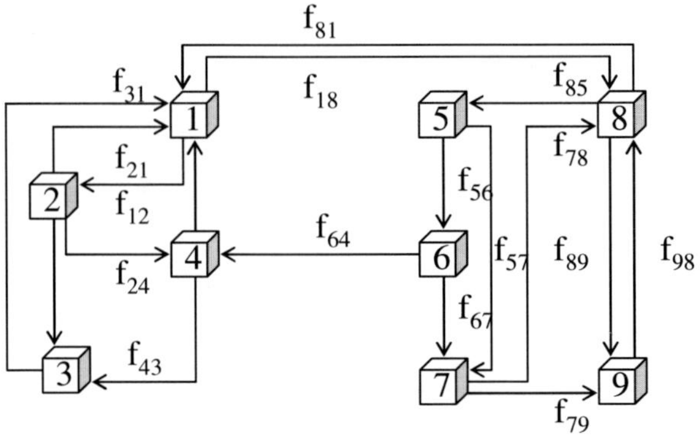

Fig. 9.8 Network of production cells

Consequently the system shown in Fig. 9.8 is characterized by the similarity matrix:

$$R = \left|\begin{array}{ccccccccc}
1 & .25 & .25 & 1 & 0 & 0 & 0 & .5 & 0 \\
 & 1 & .25 & 0 & 0 & 0 & 0 & 0 & 0 \\
 & & 1 & .5 & 0 & 0 & 0 & 0 & 0 \\
 & & & 1 & 0 & 0 & 0 & 0 & 0 \\
 & & & & 1 & 1 & .25 & .25 & 0 \\
 & & & & & 1 & 1 & 0 & 0 \\
 & & & & & & 1 & 0 & 0 \\
 & & & & & & & 1 & .25 \\
 & & & & & & & & 1
\end{array}\right| \tag{9.4}$$

Using the max-min rule of stabilization we obtained:

$$R^4 = \left|\begin{array}{ccccccccc}
1 & .25 & .5 & 1 & .25 & .25 & .25 & .5 & .25 \\
 & 1 & .25 & .25 & .25 & .25 & .25 & .25 & .25 \\
 & & 1 & .5 & .25 & .25 & .25 & .5 & .25 \\
 & & & 1 & .25 & .25 & .25 & .5 & .25 \\
 & & & & 1 & 1 & 1 & .25 & .25 \\
 & & & & & 1 & 1 & .25 & .25 \\
 & & & & & & 1 & .25 & .25 \\
 & & & & & & & 1 & .25 \\
 & & & & & & & & 1
\end{array}\right| \tag{9.5}$$

Denote by R_T the partition in classes at the level T.

We obtained for similarity degree $0.5<T1\leq 1$ the partition R_{T1} = (1,4),2,3,(5,6,7), 8,9.

For similarity degree $0.25<T2\leq 0.5$ the partition is R_{T2} = (1,3, 4,8), 2, (5,6,7),9.

For similarity degree $0<T3\leq 0.25$ the partition is R_{T3} = (1, 2, 3, 4, 5, 6, 7, 8, 9).

For T1 we are faced with 6 classes, for T2 with 4 classes and for T3 a unique class.

Examples of grouped matrices are:

$$\hat{R}_{T2} = \begin{vmatrix} 1 & .25 & .25 & .25 \\ & 1 & .25 & .25 \\ & & 1 & .25 \\ & & & 1 \end{vmatrix} \tag{9.6}$$

$$\hat{R}_{T4} = (1) \tag{9.7}$$

The corresponding entropies are: $H(\hat{R}_{T4}) = 17.65$, $H(\hat{R}_{T2}) = 6.75$ and $H(\hat{R}_{T4}) = 0$.

Grouping induces entropy decrease. A criterion for entropy decreases may impose the acceptable grouping degree.

Observe that at any considered similarity degree T, the compartments (5, 6, 7) and (1, 4) pertain to the same class. Consequently it is possible to decrease the experimental effort and limit the observation to just one element of that class.

If we consider 0.5 as an acceptable degree of similarity, we may limit the recording to 6 states only, corresponding to the classes: (1,4), 2 ,3, (5,6,7), 8, 9.

Entropy may indicate the effect of modification in the manufacturing networks. It provides a quantitative method to evaluate the performance of system layout design in terms of complexity and throughput.

References

Battini, D., Persona, A., Allesina, S.: Towards a use of network analysis: quantifying the complexity of Supply Chain Networks. International Journal of Electronic Customer Relationship Management 1(1), 75–90 (2007)

Beer, S.: Diagnosing the System. J. Wiley, Chichester (1985)

Camarinha-Matos, L.M., Afsarmanesh, H.: Collaborative networks: A new scientific discipline. J. Intelligent Manufacturing 16(4-5), 439–452 (2005)

Frei, R., Barata, J., Di Marzo Serugendo, G.: A Complexity Theory Approach to Evolvable Production Systems. In: Sapaty, P., Filipe, J. (eds.) Proceedings of the International Workshop on Multi-Agent Robotic Systems (MARS 2007), pp. 44–53. INSTICC Press, Portugal (2007)

Frizelle, G., Suhov, Y.M.: An entropic measurement of queueing behaviour in a class of manufacturing operations. Proceedings of the Royal Society London 457, 1579–1601 (2001)

Gino, F.: Complexity measures in decomposable structures. Technical report. EURAM (European Academy of Management), Conference on Innovative Research in Management, May 9-11, Stockholm, Sweden (2002)

Hu, B., Efstathiou, J.: Responsive system based on a reconfigurable structure. In: Pham, D.T., Eldukhri, E.E., Soroka, A.J. (eds.) 2nd I*PROMS Virtual Conference on Intelligent Production Machines and Systems, pp. 517–522 (2007)

Iordache, O.: Evolvable Designs of Experiments Applications for Circuits. J. Wiley VCH, Weinheim (2009)

Karp, A., Ronen, B.: Improving shop floor control: An entropy model approach. International Journal of Production Research 30(4), 923–938 (1992)

Kauffman, S.A.: The Origins Of Order: Self Organisation and Selection in Evolution. Oxford University Press, New York (1993)

Leitão, P., Restivo, F.: ADACOR: A Holonic Architecture for Agile and Adaptive Manufacturing Control. Computers in Industry 57(2), 121–130 (2006)

Lohse, N., Valtchanov, G., Ratchev, S., Onori, M., Barata, J.: Towards a Unified Assembly System Design Ontology using Protégé. In: Proceedings of the 8th Intl. Protégé Conference, Madrid, Spain (2005)

McCarthy, I.P.: Technology management ± a complex adaptive systems approach. International Journal of Technology Management 25(8), 728–745 (2003)

McCarthy, I.P.: Manufacturing Strategy—Understanding the Fitness Landscape. International Journal of Operations and Production Management 24(2), 124–150 (2004)

Mun, J., Ryu, K., Jung, M.: Self-reconfigurable software architecture: design and implementation. In: Proc. 33rd Int. Conf. Computers Industrial Engineering, Jeju, Korea, March 25-27, vol. CIE569 (2004)

Onori, M.: Evolvable Assembly Systems-A New Paradigm. In: IST 2002-33rd International Symposium on Robotics, Stockholm, pp. 617–621 (2002)

Onori, M., Barata, J., Frei, R.: Evolvable Assembly System Basic Principles. BASYS Niagara Falls, Canada (2006)

Ryu, K., Jung, M.: Agent-based fractal architecture and modeling for developing distributed manufacturing systems. Int. J. Prod. Res. 41(17), 4233–4255 (2003)

Wang, H., Efstathiou, J., Yang, J.-B.: Entropy-based Complexity Measures for Dynamic Decision Processes. Dynamics of Continuous Discrete and Impulsive Systems Series B: Applications & Algorithms 12, 829–848 (2005)

Yolles, M.I.: Organisations as Complex Systems: An Introduction to Knowledge Cybernetics. Information Age Publishing, Inc., Greenwich (2006)

Chapter 10
Concept Lattices

Abstract. Formal concept analysis for multi-dimensional data analysis is highlighted by examples.

Polyadic and temporal formal concept analyses are presented in general PSM framework.

The relation between OLAP (On-Line Analytical Processing) and lattices is outlined.

Computational biochemistry case studies are based on entropy criteria.

Emergent computing capabilities for *Physarum* systems are evaluated.

Multivariate analysis is correlated to hierarchical classes and formal concept analysis.

10.1 Galois Lattices

A lattice is a partially ordered set, also called a poset in which any two elements have a unique supremum, the elements' least upper bound, called their join, and an infimum, the greatest lower bound, called their meet. Lattices can also be characterized as algebraic structures satisfying specific axiomatic identities. Lattice theory pertains to both order theory and universal algebra.

In the area of order theory, completeness properties assert the existence of certain infima or suprema of a given poset.

An interesting way to characterize completeness properties is provided through the concept of monotone Galois connections, that is, adjunctions between partial orders. The general observation on which this reformulation of completeness is based is that the construction of suprema or infima provides left or right adjoint parts of suitable Galois connections.

Such considerations suggest a reformulation of order theory in terms of category theory, where properties are expressed by referring to the relationships as morphisms, and adjunctions between objects, instead of considering their internal structure.

Every partially ordered set can be viewed as a category in a natural way, considering that there is a unique morphism from x to y if and only if $x \leq y$. A Galois connection is a pair of adjoint functors between two categories that arise from partially ordered sets. The upper adjoint is the right adjoint while the lower adjoint is the left adjoint.

A poset may be regarded as a category in which there is at most one map between two objects.

Category theory was regarded as coherently constructive lattice theory (Backhouse and Bijsterveld 1994).

Concepts from lattice theory as preorder, monotonic function, Galois connection and closure operation are associated to category theory concepts as: category, functor, adjunction and monad, respectively. Consequently the lattice theory is a valuable source of inspiration for category theory (Lambek 1968).

The formal concept analysis, FCA, was developed as a theory of data analysis which identifies conceptual structures among datasets (Ganter and Wille 1999).

The conventional approach takes as input a matrix specifying a set of objects and the properties, called attributes, and finds both all the natural clusters of attributes and all the natural clusters of objects in the input data, where an object cluster is the set of all objects that share a common subset of attributes, and a property cluster is the set of all attributes shared by one of the natural object clusters.

Natural property clusters correspond one-for-one with natural object clusters, and a concept is a pair containing both a natural property cluster and its corresponding natural object cluster. The family of these concepts obeys the mathematical axioms defining a lattice, and is called a concept lattice or Galois lattice, GL, because the relation between the sets of concepts and attributes is a Galois connection.

The main goal of FCA has been the support of rational communication and the representation and processing of knowledge based on the so-called restructuring program (Wille 1996a, 1996b). The program of restructuring has a philosophical background which goes back to the pragmatism of Peirce. Lattice theory is reworked in order to integrate and rationalize origins, connections and interpretations in the real world.

The FCA plays a prominent role in conceptual modeling by combining the ease of handling database objects that are defined via a list of properties to a mathematical model rooted in a formalization of logic by which reasoning is based on communicative rationality in the sense of pragmatism. FCA produces graphical visualizations of the inherent structures among data.

In FCA the concept understanding is considered as the basic unit of thought. A particular concept has both an extension and an intension.

FCA supposes that some relation between objects and properties is already established in the form of a context $F = (G, M, I)$ where G is the set of objects, M is the set of properties or attributes and $I \subseteq G \times M$ is the incidence relation between objects and properties (Ganter and Wille 1999).

A formal context *F* can best be represented by a table specifying which objects fall under which properties. This suggests that a context may be associated to classification purposes.

For a set of objects $A \subseteq G$, we can define all the properties shared by all objects, provided a context *F* is given:

$$A' = \{m \in M \mid \forall g \in A: (g, m) \in I\} \quad (10.1)$$

These are the common properties of A.

Similarly the dual operation can be defined provided a property set $B \subseteq M$ is given that is:

$$B' = \{g \in G \mid \forall m \in B: (g, m) \in I\} \quad (10.2)$$

These are the common objects of B.

Assume a context *F*= (G, M, I) is given. A formal concept of *F* is defined as a pair (A, B) where $A \subseteq G$ (called extent) and $B \subseteq M$ (called intent), A′=B and B′=A.

Given a context *F* = (G, M, I) the collection of all formal concepts $\mathcal{B}$(G, M, I) forms a complete Galois lattice, GL, where the partial order ≤ is defined by (A1, B1) ≤ (A2, B2) iff $A1 \subseteq A2$ (which is equivalent to $B1 \supseteq B2$). The supremum and the infimum are defined as follows:

$$\bigvee_{t \in T} (A_t, B_t) = ((\bigcup_{t \in T} A_t)'', \bigcap_{t \in T} B_t) \quad (10.3)$$

$$\bigwedge_{t \in T} (A_t, B_t) = (\bigcap_{t \in T} A_t), (\bigcup_{t \in T} B_t)'') \quad (10.4)$$

The complete lattice $\mathcal{B}$(G, M, I) may be interpreted as a category. The operation denoted by" is a closure operator on both sides the object level, G and the attribute level, M.

The closure operation induces a complete lattice and the concept lattice corresponds to a Galois-connection between two closure operators.

Any concept lattice $\mathcal{B}$(G, M, I) determines a hierarchy of formal concepts.

This hierarchy can be used to perform inferences with respect to properties of concepts or with respect to extensions of concepts.

Table 10.1 Formal context concerning airport gates

Gates	TG	BG	DG	IG
A1-1	X	X	X	
A2-9		X	X	
A10-23	X		X	X
A22		X	X	X
B10	X		X	
B11-23		X		X
B20-9	X			X
C2	X	X		X

The illustrative example shown in Table 10.1 refers to the gates for a terminal at an airport (Stumme 1998).

The object set G comprises the group of gates denoted by A1-1, A2-9 and so on (Table 10.1).

The attributes set M four different functionalities denoted as TG-terminal gate, BG-bus gate, DG-domestic gate, IG-international gate.

Fig. 10.1 shows the GL associated to data from Table 10.1

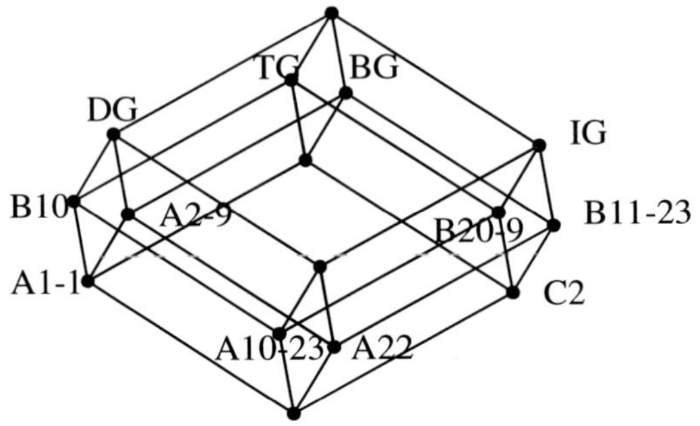

Fig. 10.1 Formal concept lattice

The visualization of line diagrams supports the study of attribute interactions.

The fact that the lattice is a 4D-cube shows that there are no implications between the four attributes that is between the four gates.

10.2 Polyadic Context Analysis

Designing efficient data mining algorithms to compute collections of relevant patterns is an active research domain. Useful knowledge discovery processes can be based on patterns extracted from large multi-dimensional datasets.

Many datasets record whether some properties hold for some objects, for instance, whether a product is bought by a customer or whether a gene is overexpressed in a biological sample. Such datasets are binary relations and can be represented as 0/1 matrices. In such matrices, a closed itemset is a maximal rectangle of "1"s modulo arbitrary permutations of the lines, that is objects, and the columns that is, properties. Thus, every closed itemset supports the discovery of a maximal subset of objects sharing the same maximal subset of properties.

Efficiently extracting every closed itemset satisfying user-defined relevancy constraints has been extensively studied.

This framework often turns out to be ineffective, first of all, since many datasets are n-ary relations, that is, 0/1 tensors. Reducing their analysis to two dimensions is ignoring potentially interesting additional dimensions, for instance, when a gene expression is measured. The presence of noise in most real-life datasets is a second issue for ineffectiveness, which leads to the fragmentation of the patterns to discover.

Effective search can be guided by the broad class of relevancy constraints the patterns must satisfy (Cerf 2010).

Conventional formal context consists of two sets, that of objects and that of attributes, together with a binary relation between objects and attributes. This relation induces in the standard way the Galois connection between sets of objects and sets of attributes whose closed sets, the formal concepts, form a complete lattice, the lattice of formal concepts. Because it employs a binary relation, a formal context is, two dimensional.

The triadic concept analysis was introduced as a generalization of formal concept analysis to three dimensions. Apart from the sets of objects and attributes, a third set, called the set of conditions, was introduced and a ternary relation between objects, attributes and conditions took the place of the binary relation of formal contexts (Wille 1995, Lehmann and Wille 1995, Dau and Wille 2001).

Complete trilattices were the lattice-theoretic structures that arose in place of complete lattices out of this generalization.

Voutsoudakis introduced the polyadic concept analysis, which generalizes triadic contexts and concepts to n dimensions for arbitrary n (Voutsoudakis 2002).

Observe that previously discussed FCA studies refer to formal contexts only, since K1, K2 and K3 are formal domains.

The self-integrative closure hypothesis requires that the formal contexts are completed with the natural or real context denoted here by S or K0.

Completing the Peirce's triadic approach, S is supposed to have a formal signification as that associated to K1, K2 and K3.

For self-integrative closure hypothesis, to any general PSM framework containing S, K1, K2 and K3 we may associate a tetradic context (S, K1, K2, K3, Y) denoted also by (K0, K1, K2, K3, Y).

A tetradic concept is the quintuple (S, K1, K2, K3, Y) where S, K1, K2 and K3 are sets and Y is a quaternary relation between S, K1, K2 and K3 that is $Y \subseteq S$ x K1 xK2 x K3. The elements of S, K1, K2 and K3 are called real states, formal objects, attributes and conditions. An element (s, k1, k2, k3) $\in$ Y is read: for the real state s=k0$\in$ K0, the object k1$\in$ K1, has the attribute k2$\in$ K2 under the condition k3$\in$ K3.

Recall that a concept is a pair of sets: a set of elements (extent) and a set of properties (intent) as (k1, k2), for example. We may define the hierarchy of context considering k1 as an extent relative to k2 and k2 as an intent relative to k1. In the same way k3 appears as intent relative to k2.

Table 10.2 shows an example of tetradic context.

Table 10.2 contains a segment of an elementary type of tetradic context. In this case: K0 = K1 = K2 = K3 = {0, 1, 2, 3}.

In this table rows represents the real objects, K0, the columns represent formal objects, K1, the subtables represent the formal attributes, K2 and the tables represents the formal conditions K3. Only the value "0" of K3 was figured.

Fig. 10.2 illustrates the tetralattice for tetradic context.

Table 10.2 Tetradic context: partial data

K3	0															
K2	0				1				2				3			
K0\K1	0	1	2	3	0	1	2	3	0	1	2	3	0	1	2	3
0		X	X	X	X	X	X	X	X	X	X	X	X	X	X	X
1	X	X	X	X	X		X	X	X	X	X	X	X	X	X	X
2	X	X	X	X	X	X	X	X	X	X		X	X	X	X	X
3	X	X	X	X	X	X	X	X	X	X	X	X	X	X	X	

Fig. 10.2 shows as points the elements of one type of Sierpinski carpet.
The complete Sierpinski carpet would correspond to a completed Table 10.2.
Visualizations for tetradic GL are complex.
A possibility is to decompose the tetradic lattices in triadic lattices.

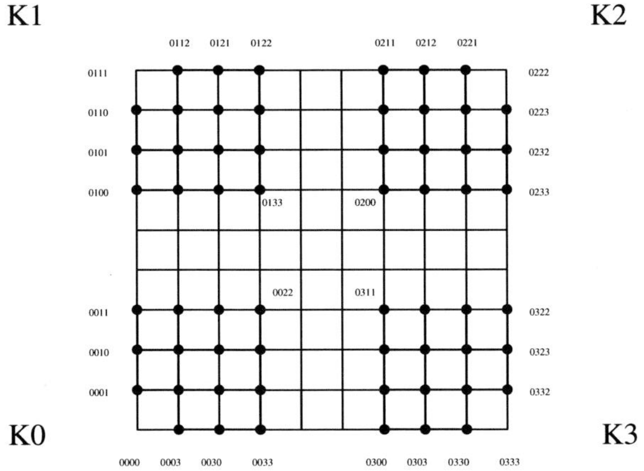

Fig. 10.2 Tetralattice for tetradic context

Fig. 10.3 outlines the different concept contexts and the self-integrative closure hypothesis. The notations are: S-Substance, K1-Firstness, K2-Secondness, and K3-Thirdness.The associated Peirce's categories are indicated.

It is considered that each new module depends and embeds the previous ones as it happens in the general categorification process.

Following categorification way a decategorification way should be considered too.

Fig. 10.3 shows that after the integration, or direct epistemology way S→K1→K2→K3 we need to look at the differentiation, or reverse epistemology way K3′→K2′→K1′→S′.

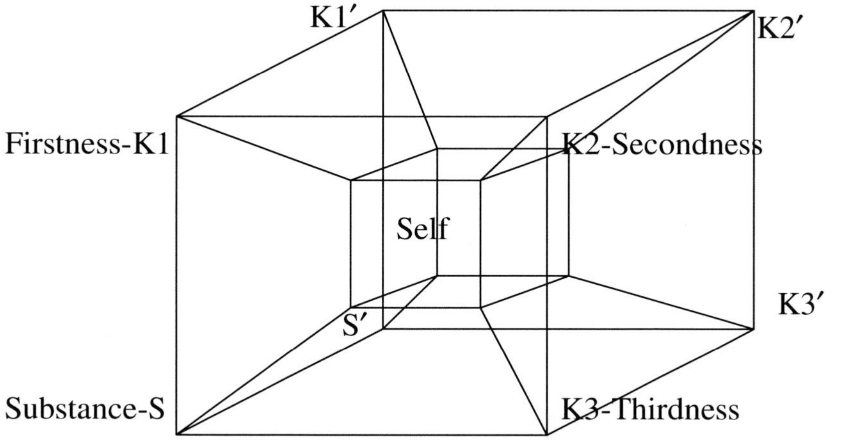

Fig. 10.3 Polytope for self-integrative closure of tetradic lattice

It was observed that making use of the developments of the direct way may offer in a kind of symmetry-breaking results for the reverse way. An account of this, the swinging from direct to reverse epistemology will be critical since the boundaries where new information is created consist of coexisting tendencies of integration and differentiation.

Fig. 10.4 shows an operad associated to the tetradic Galois lattice.

Here k0, k1, k2 and k3 select elements from the basic modules K0, K1, K2 and K3.

Observe that K3 involves three-dimensional lattices, a representation for triadic concept analysis.

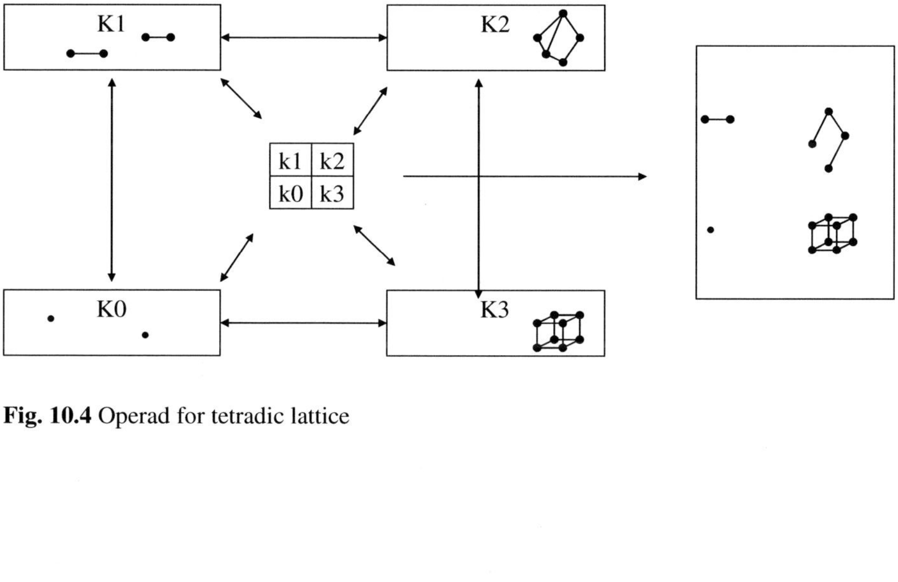

Fig. 10.4 Operad for tetradic lattice

10.3 Temporal Concept Analysis

Temporal concept analysis, TCA, is an extension of FCA in which the evolutions of the system or object are considered in conjunction with the conceptual aspects of the object. Significant researches in the area, due to Wolff (Wolff 2001, 2005, 2011) and to Wollbold (Wollbold 2007, Wollbold et al. 2008, Wollbold et al. 2011), approached the problem by adding directed edges to the lattice to capture the evolutionary behaviors of the attributes. Wollbold has focused on an FCA modeling of temporal transitions.

Wolff's studies, more directed toward a description of temporal concepts than toward temporal logic, have allowed a formal representation of the temporal extensions of FCA. Wolff has approached temporal concept analysis by scaling the time and event space and adding directed edges to the concept lattice of the context. The potential difficulty of this approach is related to the complexity of the display. Complex information bases may rapidly overwhelm an advantage lattice representation bring to formal concept analyses.

A way around this complexity issue is to redefine how we think about systems / objects and the states of those systems. Traditionally, we view a system in a specific state as a unique object, so we are forced in an FCA paradigm to replicate an object as many times as we have states for it. The system view as being unique with sets of constant or time dependent attributes allows reducing the complexity of the lattice.

We can use FCA and focus on a fixed time only.

Fig. 10.5 illustrates the temporal concept analysis.

The integration way and the differentiation way are presented. The swinging effects of this transition between the front and the back face of the external cube in Fig. 10.5 could be considerable. The use of the developments of the direct way induces new results because the boundaries where new information is created consist of parallel tendencies. Tendencies to integrate coexist with tendencies to differentiate and it is the timing and blend of both that counts for self-evolvability.

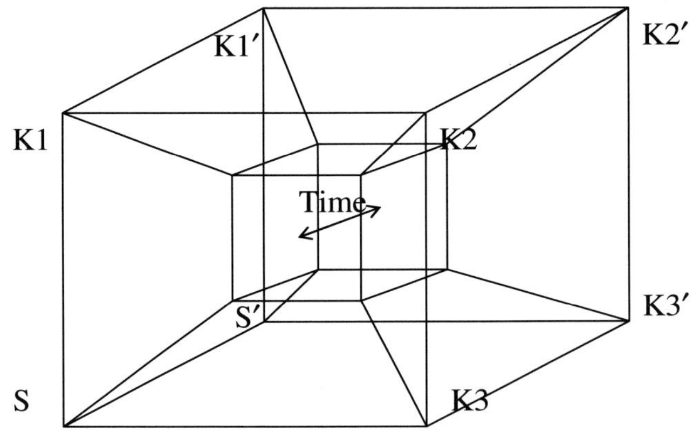

Fig. 10.5 Polytope for temporal concept analysis

10.4 OLAP and Lattices

OLAP relies on the image of a high-dimensional cube containing data (Berson and Smith 1997).

The basic idea is that datasets are viewed as cubes with hierarchies along each axis. To navigate the cube, we specify an aggregation function to say how to summarize information about groups of cells within the cube. This structure allows us to view large datasets in a highly summarized form and then expand the data along each axis as needed to see finer grained features.

One might for instance want to structure sales fact along the dimensions region, product and time. These dimensions span a data cube.

The cube is composed of cells, one for each combination of region, product and time.

The analyst may ask queries that refer to one-dimensional or two-dimensional data.

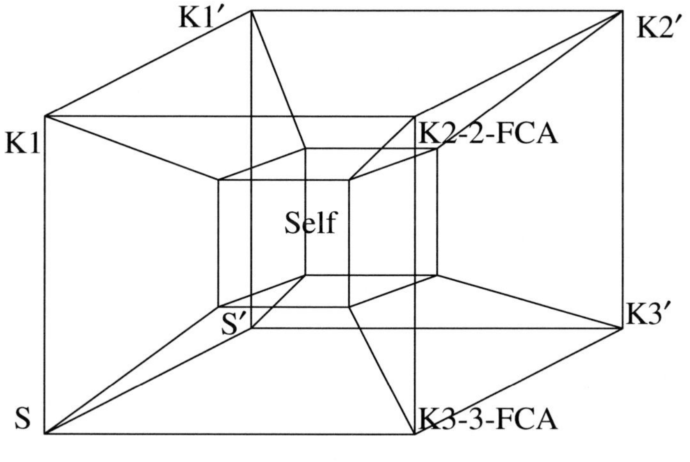

Fig. 10.6 Polytope for formal concept analysis

The reduction of dimensionality is known as slicing in OLAP.

Another capability is aggregation function corresponding to summing up the numbers.

Usually, there are predefined hierarchies on the dimensions along which the aggregation takes place.

An additional feature of OLAP is dicing, which rotates the data cube. This allows permuting rows and columns.

There exists significant correlation between lattices and OLAP (On-Line Analytical Processing) (Dehne et al. 2002, Stumme 2005)

OLAP data cube with three dimensions (G, M, B) may be considered as triadic FCA while a four-dimensional OLAP, (K0, K1, K2, K3) may be correlated to tetradic FCA and so on.

As in OLAP, FCA users should be able to dice. This means to allow using any of the three sets as the set of objects at some points in time depending on the task at hand. Which of the sets is considered as object set, attribute set, and condition set depends on the user.

Fig. 10.6 shows the polytope for FCA.

The standard FCA is denoted as 2-FCA. It produces the 2-clusters.

The 3-FCA produces the 3-clusters and corresponds to triadic context analysis.

There exist an integration way and a differentiation way. The front face of the polytope shown in Fig. 10.6 corresponds to integration while the back face corresponds to a differentiation. The reconciliation of these two ways is a process moderated by the Self.

A common representation of the data cuboids that captures the computational dependences among different groups by queries is a lattice.

Fig. 10.7 shows an example of such four-dimensional lattice (Dehne et al. 2002)

The lattice from Fig. 10.7 corresponds to a four-dimensional data cube with dimensions A, B, C and D. Every node in the lattice represents a group by query and is labeled with its grouping attributes which consists of the subset of dimensions that participate in the group by clause in the corresponding query.

Fig. 10.7 suggests a relation between such lattices and statistical factor analysis, ANOVA and Hasse diagrams.

Notice that scientific OLAP as presented by Huyn includes several capabilities related to statistical analysis (Huyn 2001).

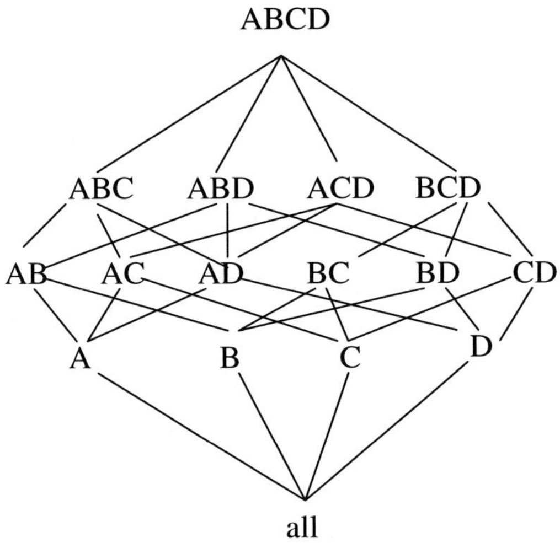

Fig. 10.7 Four-dimensional lattice

A common question in experimental studies is whether or not several groups of observations differ with respect to some measures in any significant way and not by the chance.

A statistical test commonly used to measure significance is the standard ANOVA statistic can be implemented in OLAP using an aggregate query nested within another.

Other directions of application are bases on dimensionality reducing and feature elimination.

This means to select rows and columns to retain for modeling accuracy.

Feature elimination should be based on entropic criteria. A potential method is based on informational distance DD (eq. 4.5). This distance is similar to the cross-entropy and to Kullback-Leibler divergence. Based on DD calculus a dimension may be removed from consideration if the information loss caused by the removal is small.

Fig. 10.8 shows an example of information network in multi-level datasets, based on OLAP and FCA. Similar situations have been discussed by Alqadah (Alqadah and Bhatnagar 2008, Alqadah 2010).

The levels indicated in Fig. 10.8 are S-Phenotypes, K1-Bio-processes, K2-Diseases, and K3-Drugs. In this case study the Self is associated to Genes (Deshmukh 2008).

As shown in Fig. 10.8 the genes appear as bioinformational living proofs of the Self.

The choice of modules is determined by biophysical considerations. Clustering provides significant insights into the structures embedded in the dataset.

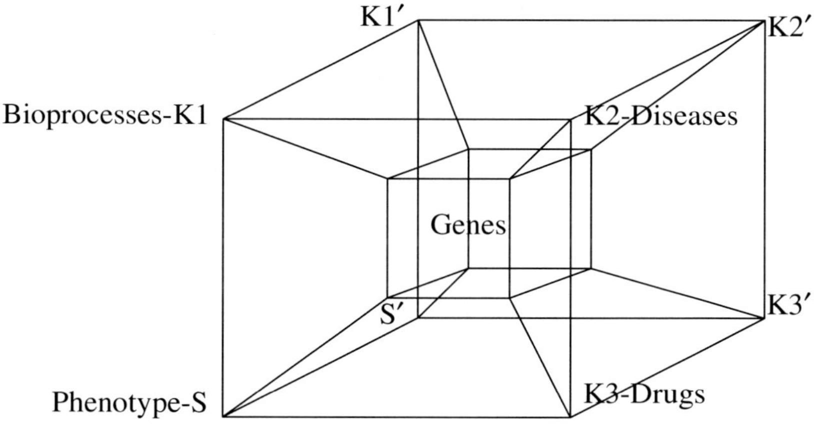

Fig. 10.8 Multi-level information network

High-dimensional and sparse datasets have been dealt with using subspace clustering algorithms to determine interesting concepts in the data. The situation of multi-level datasets, each relating two of the levels, represents a challenge. For example, we may have a gene-diseases dataset, a gene-drug dataset, a drug-phenotype adverse reactions dataset, and a gene-bio-processes dataset. We would like to discover subsets from multiple domains, in the form of a multi-level complex, that are related to each other and provide interesting insights into the datasets.

Most of the above applications listed above also comprise of multiple, interrelated datasets involving several distinct attribute sets (Alqadah 2010).

Extracting useful knowledge from such multi-level data collections requires taking the subspace clustering algorithms farther. A multi-level data collection may be viewed as an information network in which the nodes of the network are the domains and the edges are the relations between the domains. Therefore every edge of the information network represents a dataset.

For example, in the bioinformatics field we may have the domains of genes, diseases, and drugs that form the nodes of the information network. Datasets relating diseases to genes, diseases to drugs, and drugs to drugs form a meaningful collection for integrated analysis, and the resulting information network is shown in Fig. 10.8. The information network no longer entails viewing data as objects defined by feature vectors, but rather data objects from one dataset related to data objects in another dataset through shared domains.

Hence, subspace clustering in an information network entails clustering related objects across all datasets, and defining them in terms of the subsets of attributes selected from multiple domains.

Clustering across an information network is also referred to as multi-way and relational clustering. Multi-way clustering across information networks advances knowledge discovery capabilities in two manners. First, hidden associations among objects from different domains are unveiled, leading to a better understanding of the structures hidden across the entire information network.

Considering the sample information network in Fig. 10.8, hidden associations between drugs and genes may be revealed through a multi-way clustering.

We see that this information network cluster reveals associations between a set of genes and a set drugs as mediated by a set of diseases, even though, no explicit relational data are available linking genes and drugs.

In addition to revealing hidden associations, multi-way clustering sharpens and improves the local clustering of objects within a single domain by incorporating additional information via the information network.

Entropy criteria may be based on entropy definition based on similarity measures for FCA (Alqadah 2010).

10.5 Computational Biochemistry

Microarray technologies, which can measure tens of thousands of gene expression values simultaneously in a single experiment, across different conditions and over time, have been widely used in biomedical research. They have found many applications, such as classification of tumors, assigning functions to genes and grouping genes into functional pathways (Choi et al. 2006). A large collection of database is available in the public domain and several methods have been proposed for analyzing these datasets to gain biological insights. A main method for analyzing these microarray data is based on clustering, which groups set of genes, and/or groups of experimental conditions, that exhibit similar expression patterns. These include single clustering algorithms, such as hierarchical clustering, k-means, self-organizing map algorithms and biclustering algorithms. However, the challenge to derive useful knowledge from microarray data still remains.

The method of FCA builds a Galois lattice, GL, from the experimental data together with additional biological information. Each vertex of the lattice corresponds to a subset of genes that are grouped together according to their expression values and some biological information related to gene function. The lattice structure of these gene sets might reflect biological relationships in the dataset. Similarities and differences between experiments can then be investigated by comparing their corresponding lattices according to various graph measures. In the high level description, our method consists of the following main steps:

- Build a binary relation that is, a cross-table, for each experiment. The objects of the binary relation are genes; and there are two types of attributes: gene expression attributes and biological attributes. The gene expression attributes are obtained by a discretization procedure on gene expression values. The biological attributes can be any biological properties related to gene function.
- Construct a Galois lattice for each experiment's binary relation using the efficient Galois lattice algorithms
- Define a distance measure and compare the lattices. Note that the biological attributes of genes are invariant/constant for all experiments and they can be preprocessed. The ability to integrate these constant biological attributes is one of the advantages of our method over clustering methods. This is because the constant information will be canceled out in clustering methods and thus do not add any contributions.

The applicability of FCA for microarray data comparison was studied by Potter (Potter 2005). The goal in this study was to extract local patterns in the microarray data and no biological attributes were employed.

Denote by G1 to G8 the significant genes and by P1, P2 and P3 the properties. P1 means less than 5, P2 between 5 and 10, and P3 between 10 and 20 (Potter 2005).

The datasets are presented by Table 10.3 and Table 10.4. CE corresponds to context and EV to expression value.

Table 10.3 shows the gene expression dataset for the context CE1.

Table 10.3 Gene expression dataset, CE1

-	G1	G2	G3	G4	G5	G6	G7	G8
EV	3.8	15.6	8.7	2.1	3.3	7.8	14.2	2.8

Table 10.4 shows the gene expression dataset for the context CE1.

Table 10.4 Gene expression dataset, CE2

-	G1	G2	G3	G4	G5	G6	G7	G8
EV	3.8	1.6	8.7	12.1	3.3	7.8	14.2	2.8

Fig. 10.9 and Fig. 10.10 contain the expression content.

CE1	P1	P2	P3
G1	X		
G2			X
G3		X	
G4	X		
G5	X		
G6		X	
G7			X
G8	X		

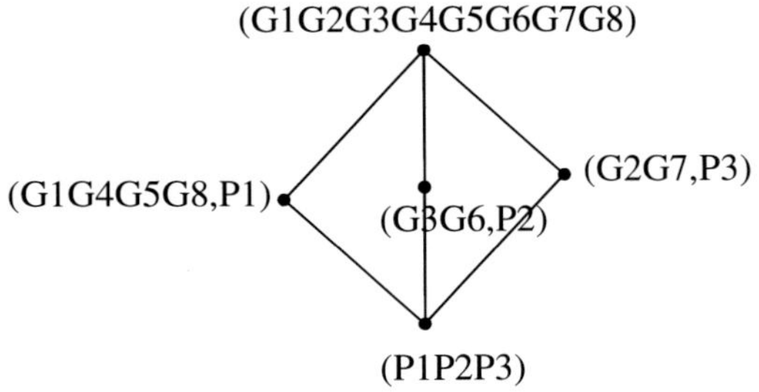

Fig. 10.9 The expression content CE1 and its Galois lattice

An open problem is to evaluate information associated to different lattices and distances between lattices.

There exist some methods based on entropy.

To the matrices shown in Fig. 10.9 and Fig. 10.10, we can associate the informational entropies: H (CE1)=29.57 and H(CE2)=30.46.

These are calculated using eq. 4.4.

Entropy calculus based on informational distance DD (eq. 4.5) for genes expression may be considered too.

CE2	P1	P2	P3
G1	X		
G2	X		
G3		X	
G4			X
G5	X		
G6		X	
G7			X
G8		X	

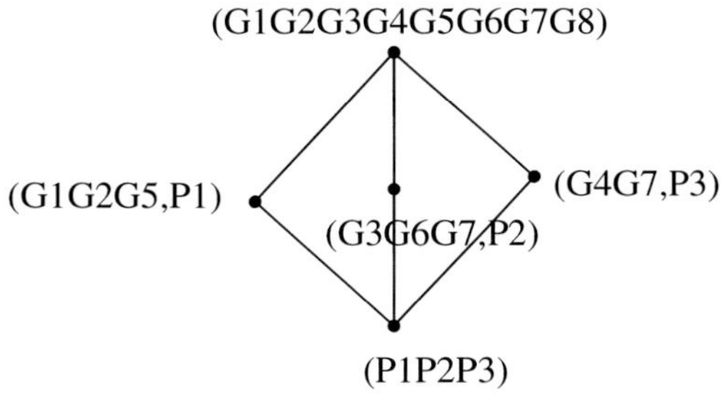

Fig. 10.10 The expression content CE2 and its Galois lattice

The procedure is to associate matrices of similarities to different expression contents and then calculate informational distance DD between the matrices.

This will show if DD is correlated to the experimental context.

A distance between graphs associated to GL may be of interest too (Potter 2005, Choi et al. 2006).

Edit distance is a graph measure, which computes the number of edges and vertices that must be added or deleted to or from one graph to be transformed into another. Edit distance is the graph analog of the Hamming distance between two strings, that is, the number of positions that differ between two strings of characters.

More formally, we define a graph G as an ordered pair $\Gamma = (V, E)$ where

V is a set of nodes (vertices) and E is a set of edges (links).

For graphs $\Gamma 1 = (V1; E1)$ and $\Gamma 2 = (V2; E2)$, the edit distance, ED, between them is given by:

$$ED (\Gamma 1; \Gamma 2) = |V1|+|V2|-2|V1 \cap V2|+|E1|+|E2|- 2|E1 \cap E2| \qquad (10.5)$$

The edit distance, ED, is therefore fairly straightforward to compute.

One could also employ a weighted edit distance for graph comparisons in which the expense for adding or removing an edge or vertex is weighted by some scalar

Both of the above-defined measures count the number of additions or deletions necessary two make the graphs equal. Edit distance, ED, can be modified further to count how many moves it would require to make two graphs close to being equal. For example, since the vertices of our graphs are labeled by genes, we can determine two vertices to be close enough if their intersection is greater than some threshold.

10.6 Emergent *Physarum* Computing Systems

FCA was applied to describe emergent *Physarum* computing systems (Tsuda et al. 2004).

The logical gates with plasmodium are designed, based on the relation between the three control parameters: (1) the gradients of the attractant, (2) the existence of other individuals, (3) free space; and the corresponding behaviors. The relation is called the context-relation expressed as Table 10.5.

Table 10.5 outlines the parameters and the behavior.

It shows that all possible combinations of three parameters, presence or absence can be discriminated in terms of the corresponding behaviors. It also means that plasmodium can recognize each combination of three parameters with respect to presence or absence. From this relation, one can estimate the logic of plasmodium in terms of lattice theory.

The logic of plasmodium is consistent with the notion of topological space that is a kind of filter to observe all possible combinations. What we take three possible parameters is chosen as a set of observable elements. A topological space is defined as a filter by which some combinations of observable elements can be observed, and explicitly corresponds to a lattice or logic. In other words, empirical data are interpreted into logic by identifiable combinations. Logic is defined not just by three control parameters but by distinction among combinations of control parameters.

With respect to the relationship between observed elements and an identifiable phenomenon resulting from combinations of observed elements, one is always employed to a particular logic that is not necessarily Boolean logic based on a set theory (Tsuda et al. 2004).

Table 10.5 Parameters and behavior

1-Gradient	2-Others	3-Space to avoid	Behavior
0	0	0	Immobilized
1	0	0	Free expansion
0	1	0	Simple avoidance
1	1	0	Simple fusion
0	0	1	Backward move
1	0	1	Forward move
0	1	1	Gradient-oriented fusion
1	1	1	Gradient-oriented avoidance

Given a set G = {a (gradient), b (existence of other plasmodium), c (presence of escape-route)}, one can obtains a power set of G, P (G), that consists of all subsets of G. If observed subsets are chosen from P (G) as S (that is S $\subseteq$ P (G)), one can obtain a topology of recognized space in the form of a lattice. The lattice is defined as a partially ordered set closed with respect to intersection and union. For example, if $S = \{\varphi, G\}$, $\varphi \cap \varphi = \varphi \in S$, $\varphi \cap G = \varphi \in S$, $G \cap G \in S$.

Then it is closed with respect to intersection and it is also verified that S is closed with respect to union, that is, for all x, y in S, $x \cup y \in S$. By contrast, if a subset {a} is also empirically observed and one obtains $S = \{\varphi, G, \{a\}\}$, then a lattice becomes a non-complemented Heyting algebra.

In the case of plasmodium, it results in S = P (G) because all combinations of three control parameters can be distinguished with each other in a term of plasmodium moves. The plasmodium's behavior depends on three parameters. The experimental condition with {a} means that there is a gradient of glucose, and {a,b} means that there is a gradient and other plasmodium individual. In both conditions there is no space to avoid other plasmodium. The motion under {a, b} is observed as simple fusion, and the motion under {a} is observed as free expansion along gradient. As a result, one can distinguish {a} from {a, b} in a term of plasmodium behavior. In analogous manners, we can distinguish all combinations of three control parameters in a term of plasmodium moves and then we can obtain S = P (G) that is a set lattice.

A lattice obtained from the context-relation is constructed as a Galois lattice (Gunji et al. 2002).

If a partial order is drawn as a line, the structure among concepts is expressed as a Hasse diagram representing finite lattice. In the described experiments, G is expressed as a set of stimulus or environmental factors G= {a (gradient), b (presence of other plasmodium), c (presence of escape-route)} and M is defined as a set of behaviors M={1 (fusion), 2 (chemotaxis), 3 (avoidance)}.

We can consider these experimental results in the context of emergent computation.

The plasmodium is adopted as computing agent because the relation between machine and its user is embedded in the form of the relationship between parts, that is local behaviors, and whole, that is global behaviors.

A *Physarum* computer shows self-repairing capabilities and could lead to emergent computing (Tsuda et al. 2004, Adamatzky 2007).

10.7 Hierarchical Classes Analysis

Efficient methods to analyze polyadic contexts may be based on hierarchical class analysis, HCA, coupled to FCA.

Both methodologies, FCA and HCA, have been applied to factorization of data for dyadic and triadic systems (Chen and Yao 2005, Hwang and Kang 2007, Belohlavek and Vychodil 2010)).

The triadic version of HCA was applied in different domains as for instance chemistry and psychiatry (Ceulemans et al. 2003).

To illustrate HCA we consider the dyadic formal context shown in Table 10.6.

Table 10.6 shows a dyadic formal context.

Table 10.6 Dyadic formal context

K1\K2	a	b	c	d	e
1	0	1	1	1	0
2	1	1	1	1	0
3	1	1	1	1	0
4	1	0	0	1	0
5	0	1	1	1	1
6	0	0	0	1	1
7	1	1	1	1	1

Table 10.6 corresponds to the objects K1 = {1, 2,...,7}, to the attributes or properties K2 = {a, b, c, d, e} and to some relations between them.

For the set G ⊆ K1 of objects and a set M ⊆ K2 of properties, two derivation operators, intent and extent, are given by Int (G) and Ext (M). Intuitively, Int (G) is the set of properties common to all objects in G ⊆ K1.

Dually, Ext (M) is the set of objects that have all the attributes from M ⊆ K2

For example Int ({1, 2, 3})={b,c,d} and Ext ({b,c,d})={1, 2,3,5,7}.

Given two objects 1, 2∈ K1, 1 and 2 are equivalent if Int ({1}) = Int ({2}).

Correspondingly an equivalence relation may be established in K2 considering the definition: a and b are equivalent if Ext ({a}) = Ext ({b}).

The hierarchical structures of the classes from Table 10.6 are shown in Fig. 10.11.

Fig. 10.11 highlights the hierarchical structure of classes for dyadic context.

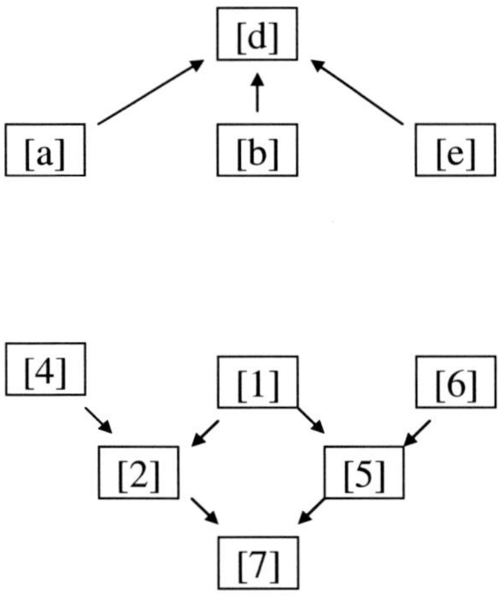

Fig. 10.11 Hierarchical structure of classes for dyadic context

An open problem is to visualize the triadic context situations.
We may consider the data from Table 10.7, as an example of triadic context.
In this case K1 = {1, 2,..,7}, K2 = {a, b, c,d,e}, and K3 = {A, B, C}.
The new considered level corresponds to conditions.

Table 10.7 Triadic context

K3	A					B					C				
K1\K2	a	b	c	d	e	a	b	c	d	e	a	b	c	d	e
1	0	1	1	1	0	0	1	1	1	0	0	1	1	1	0
2	0	1	1	1	0	1	1	1	1	0	1	1	1	1	0
3	0	1	1	1	0	1	1	1	1	0	1	1	1	1	0
4	0	0	0	0	0	1	0	0	1	0	1	0	0	1	0
5	0	1	1	1	1	0	1	1	1	1	0	1	1	1	0
6	0	0	0	1	1	0	0	0	1	1	0	0	0	0	0
7	1	1	1	1	1	1	1	1	1	1	1	1	1	1	1

In a separation schema case study A, B, C may correspond to ambient conditions as changed during the day, for instance. Variations in K3 may change the properties values and the ranking of component for different properties.

Fig. 10.12 shows how the K1 object hierarchical classification is related to K2 properties hierarchy.

The connection is mediated by K3.

Fig. 10.12 outlines the triadic class hierarchy. It shows that there exists a path from component class [2] to properties class [a] via the conditions [B], [C]. K3 intermediates between K1 and K2 allowing the closure between K1, K2 and K3.

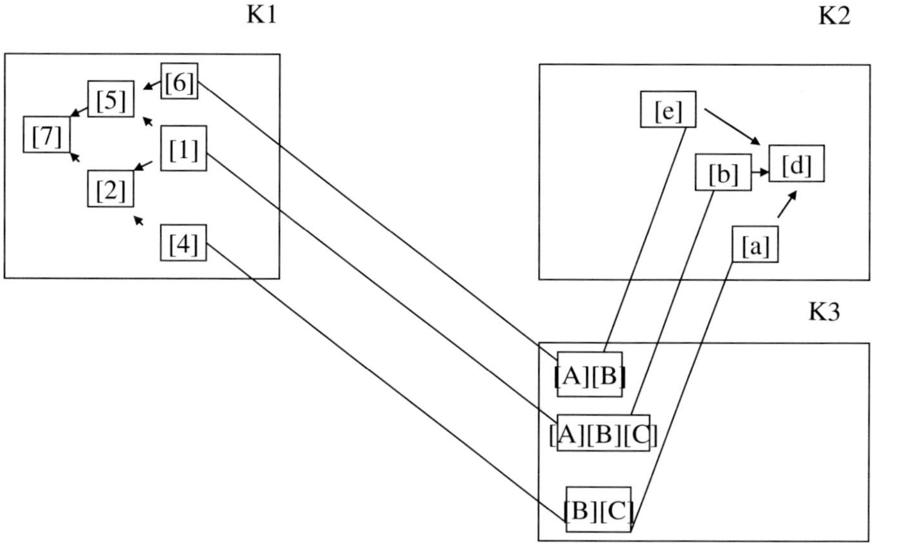

Fig. 10.12 Triadic classes hierarchy study

The hierarchical classes models are a collection of structural models for binary N-way, N-mode data (Ceulemans et al. 2003, Van Mechelen et al. 2007).

Models have been developed to study structural relation in a two-mode array and three-mode variant with binary cell entries.

Depending on the application the first mode, K1, may refer to objects, the second mode ,K2, may refer to attributes and the third mode, K3, may refer to conditions.

Table 10.8 shows a hypothetical Tucker 3-HICLAS model array of data (Ceulemans et al. 2003).

Table 10.8 Hypothetical Tucker 3-HICLAS model array

K3	A					B					C				
K1\K2	a	b	c	d	e	a	b	c	d	e	a	b	c	d	e
1	1	0	0	1	1	0	1	1	1	0	1	1	1	1	1
2	0	0	0	0	0	0	0	0	0	0	0	0	0	0	0
3	0	0	0	0	0	1	0	0	1	1	1	0	0	1	1
4	0	1	1	1	0	0	0	0	0	0	0	1	1	1	0
5	1	0	0	1	1	1	1	1	1	1	1	1	1	1	1
6	1	1	1	1	1	0	1	1	1	0	1	1	1	1	1
7	0	0	0	0	0	1	0	0	1	1	1	0	0	1	1

In this case the objects are K1 = {1, 2,..,7}, the attributes are K2 = {a, b, c, d, e}, and the conditions are K3 = {A, B, C}.

The method considers clusters called bundles of objects, attributes, conditions.

Table 10.9 shows Tucker 3-HICLAS model.

The significance is that the object bundle OB1 contains the objects {3, 5, 7} and so on.

We will have two attribute bundles, AB1 which separate according to a, d and e, and AB2 which separate according to b, c, and d. The conditions are A, B and C corresponding for instance, to three shifts A, B, C corresponding in our example to different temperatures (morning, evening, night).

Table 10.9 Tucker 3-HICLAS model

	Bundle matrices									Core array			
	Object Bundles			Attribute Bundles			Condition Bundles					Condition Bundles	
Objects	OB1	OB2	OB3	Attrib	AB1	AB2	Cond	CB1	CB2	Object Bund	Attrib Bund	CB1	CB2
1	0	1	0	a	1	0	A	0	1	OB1	AB1	1	0
2	0	0	0	b	0	1	B	1	0	OB1	AB2	0	0
3	1	0	0	c	0	1	C	1	1	OB2	AB1	0	1
4	0	0	1	d	1	1				OB2	AB2	1	0
5	1	1	0	e	1	0				OB3	AB1	0	0
6	0	1	1							OB3	AB2	0	1
7	1	0	0										

One set of conditions is CB1 that shows functioning during the shift B and C.

Core array indicates the linking structure among the three hierarchies.

For the model shown in Table 10.9 it can be derived that the Object 4 is associated with the attribute b according to the condition A, because OB3, AB2 and CB2 to which the three elements belong, respectively, are associated in the core array.

The Tucker 3-HICLAS model has a graphical representation shown in Fig. 10.13 (Ceulemans et al. 2003). Fig. 10.13 contains the overall information shown by the associated Table 10.9.

It shows that there exist some paths from object bundles, to attributes bundles via specific conditions.

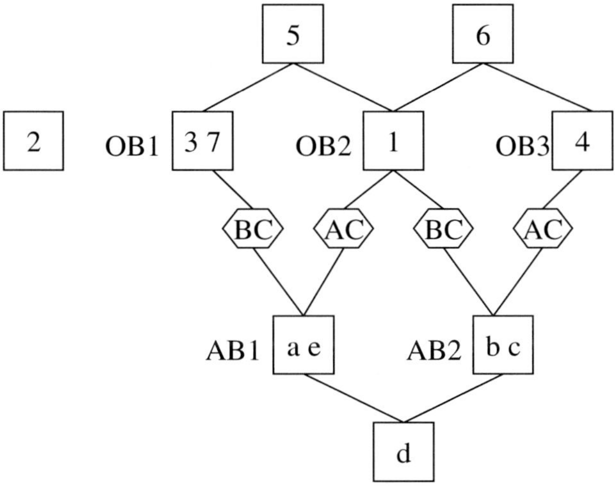

Fig. 10.13 Graphical representation of the Tucker 3-HICLAS model

The polytope from Fig. 10.14 highlights a possible development of the Tucker 3-HICLAS model.

It includes a module for substrate, S. This may be structured in bundles as the other modules of the polytope. In the example considered here, p, q, r and s are elements of the substrate, and q is present in the substrate bundle SB3.

K1 corresponds to objects bundles, K2 to attributes bundles and K3 to conditions bundles. The core arrays appear as patterns of the Self.

Fig. 10.14 suggests that after the integration way we need to look at the differentiation way. This may consist in changing the hierarchical orders in the matrices associated to S, K1, K2 and K3. Making use of the developments of the direct way will give new result for the reverse way. The swinging from direct to reverse investigation is beneficial for new model creation. For self-evolvability, tendencies to integrate data should coexist with tendencies to differentiate.

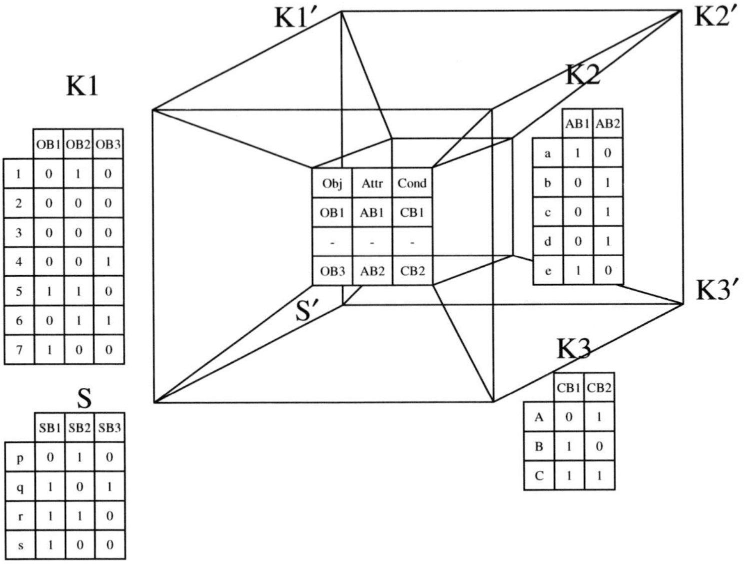

Fig. 10.14 Polytope for Tucker 3-HICLAS model

References

Adamatzky, A.: Physarum machine: implementation of a Kolmogorov-Uspensky machine on a biological substrate. Parallel Processing Letters 17, 455–467 (2007)

Alqadah, F.: Clustering of Multi-Domain Information Networks. PhD. Thesis. University of Cincinnati, Engineering: Computer Science and Engineering (2010)

Alqadah, F., Bhatnagar, R.: An effective algorithm for mining 3-clusters in vertically partitioned data. In: CIKM 2008: Proceeding of the 17th ACM Conference on Information and Knowledge Management, CIKM 2008, New York, NY, USA, pp. 1103–1112 (2008)

Backhouse, R., Bijsterveld, M.: Category Theory as Coherently Constructive Lattice Theory: An Illustration. Techical Report. Eindhoven University of Technology (1994)

Belohlavek, R., Vychodil, V.: Discovery of optimal factors in binary data via a novel method of matrix decomposition. J. Computer and System Sciences 76, 3–20 (2010)

Berson, A., Smith, S.J.: Data Warehousing, Data Mining, and OLAP. McGraw-Hill (1997)

Cerf, L.: Constraint-Based Mining of Closed Patterns in Noisy n-ary Relations. PhD. Thesis. INSA-Lyon (July 2010)

Ceulemans, E., Van Mechelen, I., Leenen, I.: Tucker 3 hierarchical classes analysis. Psychometrika 68, 413–433 (2003)

Chen, Y.H., Yao, Y.Y.: Formal concept analysis based on hierarchical class analysis. In: Proceedings of the 4th IEEE International Conference on Cognitive Informatics, ICCI 2005, pp. 285–292 (2005)

Choi, V., Huang, Y., Lam, V., Potter, D., Laubenbacher, R., Duca, K.: Using Formal Concept Analysis for Microarray Data Comparison. Advances in Bioinformatics and Computational Biology 5, 57–66 (2006)

Dau, F., Wille, R.: On the Modal Understanding of Triadic Contexts. In: Decker, R., Gaul, W. (eds.) Proc. Gesellschaft für Klassifikation, Classification and Information Processing at the Turn of the Milenium (2001)

Dehne, F., Eavis, T., Hambrusch, S., Rau-Chaplin, A.: Parallelizing the data cube. Distributed and Parallel Databases 11(2), 181–201 (2002)

Deshmukh, M.A.: Concept Based Genomic Data Exploration System. M.Sc. Thesis. University of Cincinnati, Engineering: Computer Science and Engineering (2008)

Ganter, B., Wille, R.: Formal Concept Analysis. Mathematical Foundations. Springer, Berlin (1999)

Gunji, Y.-P., Kusunoki, Y., Aono, M.: Interface of global and local semantics in a self-navigating system based on the concept lattice. Chaos, Solitons Fractals 13(2), 261–284 (2002)

Huyn, N.: Scientific OLAP for the Biotech Domain. In: Proceedings of the 27th VLDB Conference, Roma, Italy (2001)

Hwang, S.-H., Kang, Y.-K.: Applying Hierarchical Classes Analysis to Triadic context for Folksonomy Mining. In: International Conference on Convergence Information Technology, ICCIT 2006, pp. 103–109 (2007)

Lambek, J.: A fix point theorem for complete categories. Mathematische Zeitschrift 103, 151–161 (1968)

Lehmann, F., Wille, R.: A Triadic Approach to Formal Concepts Analysis. In: Ellis, G., Rich, W., Levinson, R., Sowa, J.F. (eds.) ICCS 1995. LNCS, vol. 954, pp. 32–43. Springer, Heidelberg (1995)

Potter, D.P.: A combinatorial approach to scientific exploration of gene expression data: An integrative method using Formal Concept Analysis for the comparative analysis of microarray data. Thesis dissertation. Department of Mathematics. Virginia Tech. (August 2005)

Stumme, G.: Exploring Conceptual Similarities of Objects for Analyzing Inconsistences in Relational Databases. In: Proc. Workshop on Knowledge Discovery and Data Mining, 5th Pacific Rim Intl. Conf. on Artificial Intelligence, Singapore, November 22-27, pp. 41–50 (1998)

Stumme, G.: A Finite State Model for On-Line Analytical Processing in Triadic Contexts. In: Ganter, B., Godin, R. (eds.) ICFCA 2005. LNCS (LNAI), vol. 3403, pp. 315–328. Springer, Heidelberg (2005)

Tsuda, S., Aono, M., Gunji, Y.P.: Robust and emergent Physarum logical-computing. Biosystems 73(1), 45–55 (2004)

Van Mechelen, I., Lombardi, L., Ceulemans, E.: Hierarchical classes modeling of rating data. Psychometrika 72, 475–488 (2007)

Voutsadakis, G.: Polyadic Concept Analysis. Order 19, 295–304 (2002)

Wille, R.: The basic theorem of triadic concept analysis. Order 12, 149–158 (1995)

Wille, R.: Restructuring mathematical logic: an approach based on Peirce's pragmatism. In: Ursini, A., Agliano, P. (eds.) Logic and Algebra, pp. 267–281. Marcel Dekker, New York (1996a)

Wille, R.: Conceptual structures of multicontexts. In: Eklund, P.W., Ellis, G., Mann, G. (eds.) Conceptual Structures: Knowledge Representation of Interlingua, pp. 23–39. Springer, Heidelberg (1996b)

Wollbold, J.: Attribute Exploration of Discrete Temporal Transitions. In: Kuznetsov, S.O., Schmidt, S. (eds.) ICFCA 2007. LNCS (LNAI), vol. 4390, pp. 121–130. Springer, Heidelberg (2007)

Wollbold, J., Guthke, R., Ganter, B.: Constructing a Knowledge Base for Gene Regulatory Dynamics by Formal Concept Analysis Methods. In: Horimoto, K., Regensburger, G., Rosenkranz, M., Yoshida, H. (eds.) AB 2008. LNCS, vol. 5147, pp. 230–244. Springer, Heidelberg (2008)

Wollbold, J., Huber, R., Kinne, R., Wolff, K.E.: Conceptual Representation of Gene Expression Processes. In: Wolff, K.E., Palchunov, D.E., Zagoruiko, N.G., Andelfinger, U. (eds.) KONT 2007 and KPP 2007. LNCS, vol. 6581, pp. 79–100. Springer, Heidelberg (2011)

Wolff, K.E.: Temporal Concept Analysis. In: Mephu Nguifo, E., et al. (eds.) ICCS 2001 International Workshop on Concept Lattices-Based Theory, Methods and Tools for Knowledge Discovery in Databases, pp. 91–107. Stanford University, Palo Alto, CA (2001)

Wolff, K.E.: States, Transitions, and Life Tracks in Temporal Concept Analysis. In: Ganter, B., Stumme, G., Wille, R. (eds.) Formal Concept Analysis. LNCS (LNAI), vol. 3626, pp. 127–148. Springer, Heidelberg (2005)

Wolff, K.E.: Applications of Temporal Conceptual Semantic Systems. In: Wolff, K.E., Palchunov, D.E., Zagoruiko, N.G., Andelfinger, U. (eds.) KONT 2007 and KPP 2007. LNCS, vol. 6581, pp. 59–78. Springer, Heidelberg (2011)

Chapter 11
Design of Experiments

Abstract. Latin squares and hypercubes are obtained as solutions of the wave equation.

Multivariate modeling potential for evolvable designs of experiments is evaluated.

The general PSM framework is presented as flexible guideline for a large variety of designs of experiments.

Case studies refer to pharmaceutical pipeline, to drug discovery and development and to printed circuits quality evaluations. New informational entropy criteria have been applied for 2-phenylindole derivatives library design.

11.1 Convection Model

The connection with designs of experiments was established and illustrated in the study of matrices generations flowsheeting and coupling of operations (Iordache 2009, 2010).

Consider the convective part of the first-order wave equation, WE (eq. 3.4, eq. 3.5):

$$\frac{\partial Y}{\partial T} \oplus V \otimes \frac{\partial Y}{\partial Z} = 0 \tag{11.1}$$

The initial condition is:

$$Y(Z, 0) = F(Z) \tag{11.2}$$

The operations are the sum ⊕ and the product ⊗ in GF (m).

The general solution of the partial first-order wave equation, WE, is:

$$Y(Z, T) = F(Z \oplus (V \otimes T)) \tag{11.3}$$

Consider the initial condition:

$$Y(Z, 0) = F(Z) = Z \tag{11.4}$$

This means that at T=0, the output Y of the classification schema at the distance Z in schema is exactly Z. This schema is that in which each new classification level activates a new difference in properties allowing classification. The initial condition ensures that the wave of the classification or separation process is initiated and is going on.

It results in the characteristic:

$$Y = Z \oplus (V \otimes T) \tag{11.5}$$

The GF (3) solution is presented in detail. For T=0 the solution Y is shown in Table 11.1.

Table 11.1 Convection model, m=3: Y (0, Z)

Z\V	0	1	2
0	0	0	0
1	1	1	1
2	2	2	2

For T=1 the solution is shown in Table 11.2.

Table 11.2 Convection model, m=3: Y (1, Z)

Z\V	0	1	2
0	0	1	2
1	1	2	0
2	2	0	1

For T=2 the solution of the convection model is shown in Table 11.3:

Table 11.3 Convection model, m=3: Y (2, Z)

Z\V	0	1	2
0	0	2	1
1	1	0	2
2	2	1	0

There is a relation between different solutions of the first-order wave equation, WE, and conventional DOE matrices.

For different values of T, T=1, T=2 one obtained different (3x3) Latin-squares.

Latin squares close association to DOE is well known (Hedayat et al. 1999)

The procedures to obtain DOE are suggested by universal constructions in categorical framework. There are several DOE to be obtained by combining the solutions obtained for different values of T.

Table 11.4 Concatenated solutions, m=3

000	012	021
111	120	102
222	201	210

Superposing by concatenation the elements of the Table 11.1, Table 11.2, and Table 11.3, Table 11.4 will result.

Table 11.4 shows the concatenated solutions, for m=3.

Pasting down the 3-digit numbers from Table 11.4, column after column, Table 11.5 is obtained.

Table 11.5 contains the pasting-down columns.

It is a DOE with nine experiments for three factors, denoted here F0, F1, and F3.

Table 11.5 Pasting down columns, m=3

F0	F1	F2
0	0	0
1	1	1
2	2	2
0	1	2
1	2	0
2	0	1
0	2	1
1	0	2
2	1	0

Columns in Table 11.5 are orthogonal. Each column corresponds to first-order wave equation, WE, solutions at different velocities V. Associating one supplementary digit for each column in Table 11.4, the four-digit numbers as in Table 11.6 result. Here (0) is associated to the first column in Table 11.4,(1) to the second column and (2) to the third column.

The resulting 4-digit numbers from Table 11.6 correspond to columns of well-known orthogonal design with 9 experiments and 4 factors (Taguchi 1986, 1987, Hedayat et al. 1999).

Table 11.6 shows the indexed concatenated solutions for m=3.

Table 11.6 Indexed concatenated solutions, m=3

(0)000	(1)012	(2)021
(0)111	(1)120	(2)102
(0)222	(1)201	(2)210

Concatenation and pasting-down operations are related to the coproduct "$\cup$" type of operation in categorical framework.

The previously obtained matrices are linked to the tensor product interpretation as coproduct "$\cup$". Obviously making use of tensor products as categorical product "$\times$" will give another class of solutions, asking for significantly more experiments.

11.2 Latin Hypercubes

Computer experiments are widely used for the design and development of products.

An actual reason for promoting the use of computer experiments is that physical experimentation is maybe expensive or out-of-the-way. Latin hypercube designs are beneficial for space-filling capability (Cioppa and Lucas 2007).

A class of orthogonal Latin hypercubes obtained by Ye will be presented as WE solution (Ye 1998).

The products we use, Kronecker or Hadamard, are necessary to reduce the dimensionality for data (Kolda and Bader 2009).

The construction is based on three matrices denoted by S, M and T (Ye 1998, Nguyen 2008).

The matrix S shown in Table 11.7 is obtained as solution of the kinetic part of the WE (section 3.3).

Table 11.7 Kinetic model: Matrice S

1	1	1	1
1	-1	1	-1
1	1	-1	-1
1	-1	-1	1

The matrix S is a Walsh-Hadamard design.

The matrix M shown in Table 11.8 is obtained as solution of the convection part of the wave equation WE.

Table 11.8 Convection model: Matrice M

1	2	3	4
2	1	4	3
3	4	1	2
4	3	2	1

The matrix M is a Latin square.

The matrix T as shown in Table 11.9 results using Hadamard product of S and M.

Table 11.9 Hadamard product: Matrice T

1	2	3	4
2	-1	4	-3
3	4	-1	-2
4	-3	-2	1

The Latin hypercube results by the method described in the literature (Ye 1998), Cioppa and Lucas 2007).

Table 11.10 shows the Latin hypercube.

This consists in pasting down two times the matrix T with an intercalated null row.

Table 11.10 Latin hypercube

1	2	3	4
2	-1	4	-3
3	4	-1	-2
4	-3	-2	1
0	0	0	0
-1	-2	-3	-4
-2	1	-4	3
-3	-4	1	2
-4	3	2	-1

The wave equation appears to be the source of new classes of orthogonal Latin hypercubes.

11.3 Self-Evolvable DOE Frames

The design of experiments for the exploration of high-dimensional experimental spaces may be addressed by evolvable DOE and EDOE methods.

Significant applications concern drug discovery.

We will consider that the activity of new drug discovery can be divided into four basic modules or steps.

The first module K0 corresponds to resources and to research step.

The second module K1 should be based on designs of experiments, DOE.

The third module K2 is a meta-design and for this reason it was denoted by 2-DOE.

The fourth module K3 is a meta-meta-design and for this reason may be denoted by

3-DOE.The general method is illustrated in Fig. 11.1.

The four modules of variation, K0, K1, K2 and K3, are denoted also by S, 1-DOE, 2-DOE, and 3-DOE.

To start the EDOE, we examine experimental space of properties.

After a number of iterations at this level, we may make predictions of drug-likeness too.

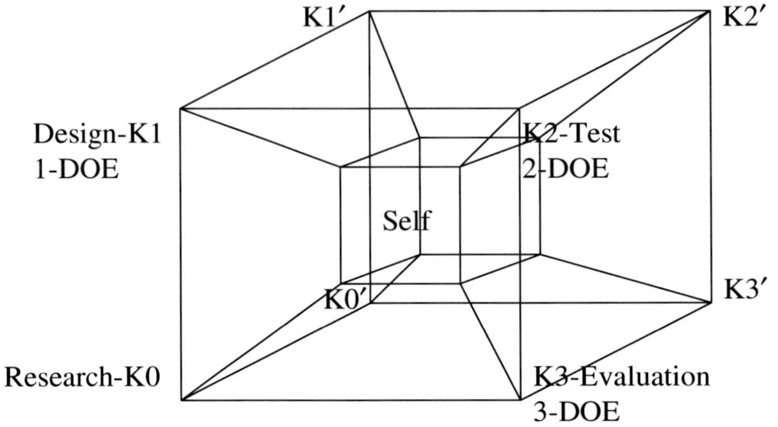

Fig. 11.1 Polytope for self-evolvable DOE basic framework

The notations are: K0-Research, K1-Design, 1-DOE, K2-Tests, 2-DOE, and K3-Evaluation, 3-DOE.

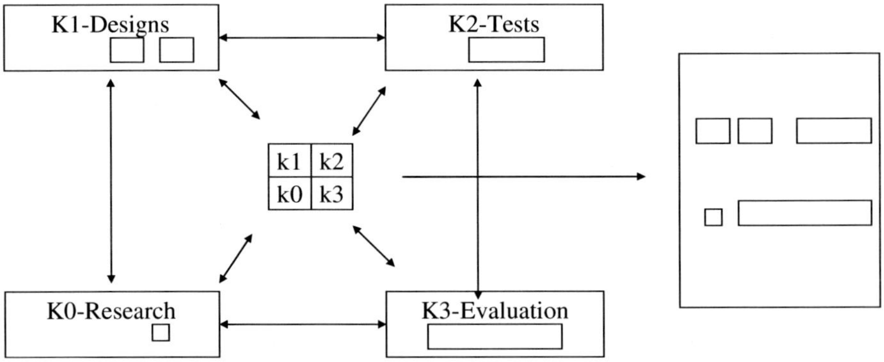

Fig. 11.2 Operad for self-evolvable DOE framework

Fig. 11.1 suggests that after the integration way we need to look at the differentiation way. This may consist in changing the hierarchical orders in the matrices associated to different DOE. Making use of the developments of the direct way may result in a kind of symmetry-breaking result for the reverse way. The swinging from direct to reverse investigation is beneficial for new designs testing because the boundaries where creative research grows and new information is created consist of synchronized integration and differentiation tendencies.

Fig. 11.2 highlights the operadic aspects of the self-evolvable DOE framework.

The DOE may be associated to a set of molecules or an embedded design if a genomic analysis is possible. The fourfold framework may be applied to just one of the levels.

Table 11.11 summarizes the categorification steps for DOE

Table 11.11 Categorification for DOE

Level	K0	K1	K2	K3	Self
-	n=0	n=1	n=2	n=3	n≥4
Categories	0-category	1-category	2-category	3-category	4-category
Example	Research	DOE	2-DOE Tests	3-DOE Evaluation	Self Evolvable

The study of DOE for self-integrative closure and the emergence of self-evolvable DOE systems corresponding to n≥4 represent a challenge.

Let us restrict here to the discovery stage associated to K1 as a first example.

Different classes of Latin square designs may be obtained as solutions of the wave equation.

A method of designing chemical substances was presented by Wood and Rose (1999).

The method allows sampling combinatorial chemistry space for synthesis based on DOE with Latin squares or more general with orthogonal arrays.

Libraries with four sites of variation for molecules may be designed using Greco-Latin squares.

Consider four sites of variation, k10, k11, k12 and k13 for substitute groups. They correspond to sub-levels of the level K1.

Then only four different substitutes are selected for each substitute or pendant group, k10, k11, k12 and k13.

The substitute group k10 consists of four candidates, denoted 1, 2, 3 and 4, the substitute k11 from four candidates denoted a, b ,c ,d, the substitute k12 of four candidates denoted A, B, C, D and substitute k13 of four candidates denoted α, β, γ, and δ.

Recall that the wave equation is able to generate Latin squares as solutions if the algebraic structures of functions and parameter are Galois Fields (Iordache 2009, 2010).

Superposition of such solutions of the wave equation gives Greco-Latin squares as shown in Table 11.12. This superposition represents a specific categorical product.

Table 11.12 shows the matrix of a Greco-Latin design.

For this table the sub-levels of the level K1 are: k10 = {1, 2, 3, 4}, k11 = {a, b, c, d}, k12 = {A, B, C, D}, and k13 = {α, β, γ, δ}.

Table 11.12 Greco-Latin square design

k10\k11	a	b	c	D
1	Aα	Bβ	Cγ	Dδ
2	Bδ	Aγ	Dβ	Cα
3	Cβ	Dα	Aδ	Bγ
4	Dγ	Cδ	Bα	Aβ

With 16 experiments only we may obtain significant information.

Running the experiment, we may select the critical substituents.

EDOE framework should be seen as a general pattern rather than as a predetermined fixed plan. This means that we may have partial Greco-Latins as micro-arrays.

For different values of time T in the solution of the convection part of the wave equation, WE, we may obtain all the orthogonal Latin squares.

A superposition of this mutually orthogonal Latin square gives the so-called Trojan squares. The presentation as Latin cube is shown in Fig. 11.3.

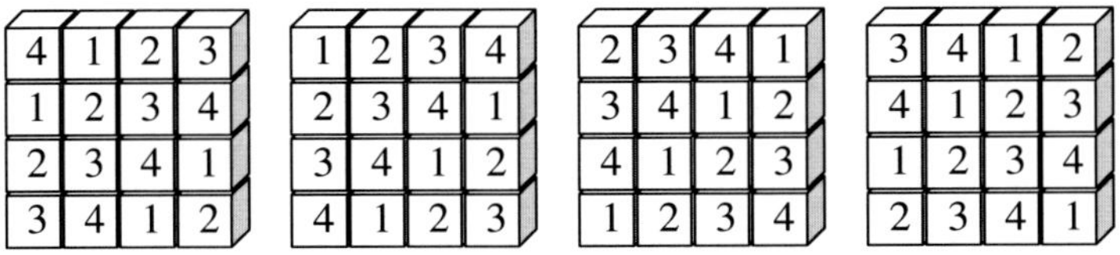

Fig. 11.3 Latin cube of order four

The wave equation may generate inflated Latin squares as solutions if the algebraic structures of functions and parameters are cyclic groups. Such modified Latin squares have been studied by Bailey (Bailey 1992). Consider that the sub-levels of K1 are the same.

The cyclic-group-based solutions of the wave equation give inflated Latin squares shown in Table 11.13. The inflation which replaces each letter by four new letters corresponds to another definition of the categorical product.

Table 11.13 Inflated Latin square design

1aAα	2bBβ	3cCγ	4dDδ
4dDδ	1aAα	2bBβ	3cCγ
3cCγ	4dDδ	1aAα	2bBβ
2bBβ	3cCγ	4dDδ	1aAα

Table 11.14 Comparison of DOE and self-evolvable DOE

Characteristics	Conventional DOE Robust design	Self-evolvable DOE
Resources	Considered very large	Definite resources-time frames
Objectives	One solution, one goal, mono-centric, search for ideal optimal solutions	General goals, multi-centric, workable, acceptable, viable solutions
Strategies	Prescribed, fixed	Self-adaptive, self-evolvable
Matrices	Fixed matrices–specific, developed Defined-based on learning Conventional design-detailed models	Constructed, self-adaptive Sensitive to case study, undefined Generated by WE and evolved
Methods	Single DOE, Sequential	Frameworks of small DOE Parallel and sequential
Required data	Clear steps, complete data	Incomplete data and variables
Factor hierarchy	Ignored; Pareto charts; inner-outer factors	Centered-heterarchy, self-evolvable frames
Approach	Mainly deductive Isolate from medium	Deductive and inductive Builds, dissociates, reconstructs
Interactions	Typically neglected, linear	Considered from start, useful
New experiments	Controlled use, confirmation experiment	Self-adaptive, evolution, minimal
Selectivity	Selective, robust to noisy	Robust to redundant information and noisy
Criteria	Average, variance, signal/noise	Informational entropy, distance
Calculus	Real field	Real and other than real fields
Data analysis	Developed-completed	Self-adaptive, interactive, online
Learning	Early-learning	Beyond learning, creative, self-evolving
Time concept	Neglects time or considers linear time only	Time scales, cyclic time frames
Complexity	Considered as enemy, avoided	Considered as ally, accepted and utilized

The complete EDOE frame implies to continue the cycle from DOE matrix that is from K1 level, to tests that is K2 level, evaluation and implementation that is K3 level as shown in Fig. 11.1.

Table 11.14 compares conventional DOE to self-evolvable DOE methods.

11.4 Multi-level Data Analysis

A structural model pertaining hierarchical classes analysis is presented in the following. The case study refers to solderability and surface finishes. Solderability is the ability of a surface to be wetted by molten solder.

Good solderability for the PCB, as well as for the components has become an important element in achieving the quality required in competitive markets. Microelectronics requires the production of reliable assemblies in an ultra-low volume environment. As the new assembly technologies, such as ball grid array, flip chip, and chip on board, have progressed, there has been a demand to obtain new solderable surface finishes, alternative to the conventional hot-air level soldering, HASL. Complex PCBs demand to increase the functionality of the final surface finish.

The challenge for printed circuits industry is to correlate the surface finish technology to specific application.

Table 11.15 shows the surface finish quality framework.

Table 11.15 includes the main factors for the surface finish quality (Iordache 2009).

They pertain to design, materials, processes and applications.

Table 11.15 Surface finish quality framework

Design, D	Materials, M	Process, P	Application, A
D1-Pad size	M1-SF type	P1-SF thickness	A1-Therm. Cycling
D2-Hole size	M2-SM type	P2-SM application	A2-IST
D3- Heat transfer	M3-Flux type	P3-Contamination	-
-	M4-Solder type	-	-

SF-surface finish, SM-solder mask, IST-interconnect stress test.

The surface finish selection system is based on EDOE methodology. The center DOE contains PCB design factors, materials factors referring to surface finish and its application for solderability, processing factors and testing factors referring to solderability after reliability tests (thermal cycling and IST). DOE steps alternate with measurements and analyze steps followed by DOE reorganization.

Different tests may be performed to evaluate surface finish, SF, solderability.

Examples of typical tests are:

- MUST- wetting balance test
- SERA- sequential electrochemical reduction analysis
- Dip & Look-standard solderability test.

The problem is to quantify the results of all these solderability tests. An example of global criterion that summarizes the partial test significance is introduced next. It associates a real value to a digitalized vector and allows subsequent modification of the DOE matrix.

Denote by S= [i_1,...,i_k...] the test resulting vector for a sample. Here i_k is the digit "1" or "0" corresponding respectively to the result pass or fail of the kth test. To any global test result S, a valuation V(S) defined by: $V(S) = \Sigma_k \, i_k \, (0.5)^k$ is associated. According to the valuation formula, the solderability for any sample depends on the significance associated in the hierarchical testing sequence to partial tests. For this case study, the first test has a weight 0.5, the next 0.25, the following 0.125 and so on, the proposed hierarchy for tests being: MUST>SERA>Dip& Look>....

The valuation V(S) is in fact a similarity as defined in Section 2.3. V(S) gives similarities relative to a reference vector containing only "1" as coordinates.

The experiment is based on EDOE methodology. DOE steps alternate with measurements and analyze steps followed by DOE reorganization.

Table 11.16 shows the element of surface finish-Design-D

Table 11.16 Surface finish-Design-D

Factor	"-1"	"1"
D1-Pad size	Small	Large
D2-Hole size	Small	Large
D3-Heat transfer	With	Without

The test mini-coupon allows performing all the solderability tests after an imposed number of interconnect stress test, IST, cycles. It evaluates solderability as a function of testing time.

Fig. 11.4 shows the modules of self-evolvable DOE frame for surface finish.

Fig. 11.4 suggests that we need to look at the integration way, K0→K1→K2→K3 and at the differentiation way, K3′→K2′→K1′→K0′.

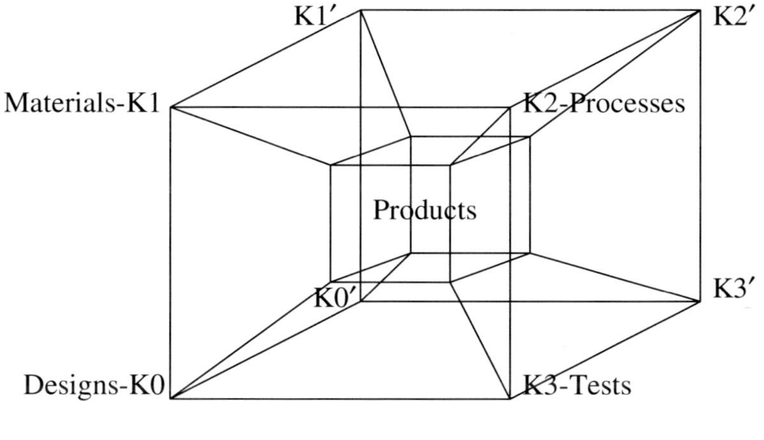

Fig. 11.4 General self-evolvable DOE frame

Preliminary tests assured that materials-M and processes-P factors are significant variables. Table 11.17 contains the notations for materials-M settings. ENIG denotes electroless nickel, immersion gold finish. S/M type depends on supplier (“D”, “C” or “T”).

Table 11.17 Surface finish-Materials-M

Factor	“-1”	“0”	“1”
M1-Surface finish	ENIG(P)	Electroplate Ni/Au	Ag
M2-Solder masks	“D”	“C”	“T”

The DOE matrix of type L 9,2,3 with three settings is considered (Iordache 2009). The main factor is M2 that is the S/M type. The processing-P factors are included in Table 11.18.

Table 11.18 Surface finish-Processing-P

Factor	“-1”	“1”
P1-Coating thickness	High	Low
P2-SM application	After	Prior

From the DOE it results that the factor P1, the coating thickness, is more significant for solderability. An interaction experiment for M, P and A was performed at this stage on the first level of EDOE. The electroless Ni/Au and a compatible mask indexed by “T” have been selected to perform an application test. Nine values of the thermal cycling parameters corresponding to the application test have been considered. The time step for the number of cycles is 25 cycles. In this case -4 is linked to 0 cycles, -3 to 25 cycles, and so on till +4 that is linked to 200 cycles. The corresponding matrices are of the type Ln,m,s (Iordache 2009).

Table 11.19 shows the factors for materials-M, processing-P, and application-A.

Table 11.19 Factors for materials-M, processing-P, application-A

<table>
<tr><td>Factor</td><td>-4</td><td>-3</td><td>-2</td><td>-1</td><td>0</td><td>1</td><td>2</td><td>3</td><td>4</td></tr>
<tr><td>A1-IST cycle</td><td>0</td><td>25</td><td>50</td><td>75</td><td>100</td><td>125</td><td>150</td><td>175</td><td>200</td></tr>
<tr><td>Factor</td><td colspan="3">“-1”</td><td colspan="3">“0”</td><td colspan="3">“1”</td></tr>
<tr><td>M2-SM type</td><td colspan="3">“D”</td><td colspan="3">“C”</td><td colspan="3">“T”</td></tr>
<tr><td>P1-Coating thick.</td><td colspan="3">High</td><td colspan="3">Avg.</td><td colspan="3">Low</td></tr>
</table>

The M, P, A factors are lumped together in the interaction type of experiment from Table 11.19. This test shows that the application factor that is thermal cycling plays significantly. The set of resulting DOE matrices is useful in the implementation of new surface finish technology.

The EDOE method allows simulations for new surface finish.

New vectors, that is, settings of parameters for an experiment may enter as new rows in the DOE matrices. New vectors have been denoted as italicized rows.

Suppose for instance that the new processing vector will be: [*1*, *-1*]. This designation is followed by a forward step. The performed experiment will be classified as run 2 in processing-P matrix. This allows predicting solderability valuation of 0.75. Consider also a new designation step in which the design-D vector is [*1*, *1*, *-1*]. This is classified in the same class as the first run in matrix D. If high solderability means valuation higher than 0.75 these two vectors that is new experiments will provide digits "1" in the center matrix. These replacements summarize the information and represent backward steps. They translate a real valuation of the solderability into a digit only. Observe that a calculus of valuation that is of similarity allows shifting from forward to backward steps.

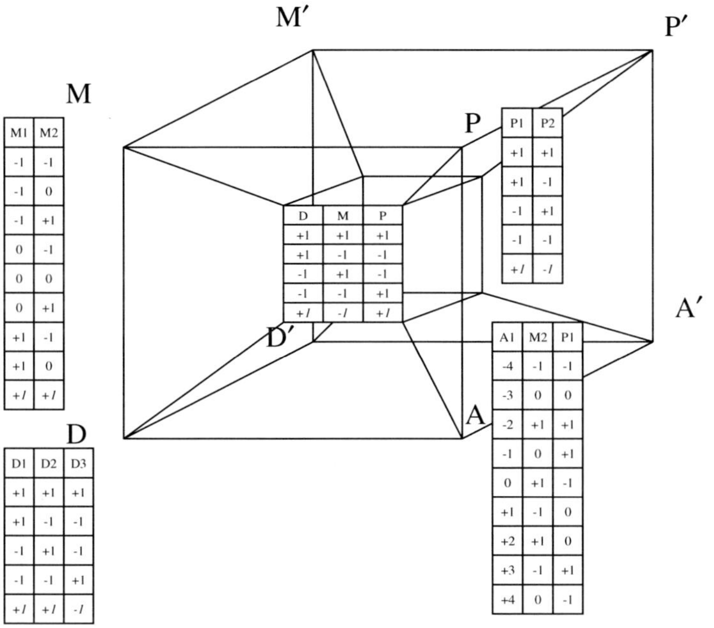

Fig. 11.5 Self-evolvable DOE frame for surface finish

Coupled with materials M corresponding to lower than 0.75 solderability (that is "-1" digit in the central matrix) the new vector in the central matrix will be [*1*, *-1*, *+1*]. This is similar to the second run [1, -1, -1] in the initial center design and it is predicted that it will show performances similar to that run.

Fig. 11.5 suggests that after the integration way D→M→P→A we need to look at the differentiation way A′→P′→M′→D′. This may consist in changing the hierarchical order or the elements in the matrices associated to D, M, P and A, in removing less significant factors. Making use of the developments of the direct way will result in a kind of symmetry-breaking result for the reverse way. The swinging from direct to reverse investigation is beneficial for self-evolution.

11.5 Pharmaceutical Systems

Designing, building and controlling complex systems became a central challenge for scientists and engineers in the coming years. A new approach to problem solving for complexity is represented by the evolvable designs of experiments, EDOE (Iordache 2009). It is based on the thesis that knowledge cannot be a passive reflection of reality, or a passive application of a formal problem-solving model, but has to be more of an active and interactive construction. EDOE is a modern way to cross industrial and technological complexity frontiers by replacing pre-programmed and fixed designs and problem-solving methods by evolvable ones.

The EDOE methodology may find applications for complex problems as the so-called pharmaceutical pipeline.

This refers to the new product, to research and development in pharmaceutical industry.

The typical sequence for new product implementation contains the following main steps:

Resources →Discovery→ Development→ Launching

Biological, chemical and other resources allow the discovery of drug lead.

The development step includes tests, preclinical, P0, followed by three phases of tests, PI, PII, and PIII.

The product launching starts with NDA, New Drug Application, and FDA, Food and Drug Administration, submissions and reviews, and continues with production and marketing steps.

Some areas of pharmaceutical industry are facing a productivity crisis (Woodcock and Woosly 2008). Despite rising investment in pharmaceutical research and development, successful development of new drugs is slowing. The high costs of new drugs development may discourage investment in more innovative, risky approaches in therapeutics.

The FDA, with its dual role of promoting and protecting health is charged with implementing policies that ensure that the benefits of the new products will surpass their risks, while simultaneous by promoting innovations that can improve health.

It was observed that chemical and biological systems may have huge behavior spaces and laboratory experiments and models cover only tiny aspects of a system's behavior.

The models often ignore the essential temporal and conceptual space organization of the research and implementation components. Moreover, models and methodologies lack flexibility to adapt and to faster represent more areas of the behavior space.

They neglect synergies – beneficial, nonlinear interactions between systems that cannot be inferred from existing resources and may be missed.

The architecture of the models should be in correspondence with that of the studied system within physically, biologically or cognitive recognizable spaces.

This will require combining multiple-level modeling methods in innovative ways, multiple levels of organization activated both in parallel as in series.

It is a need for new modeling and simulation methods, sufficiently flexible, adaptable and evolvable that is able to explore larger portions of the behavior space, a strong request for cognitive architecture reflecting the essential temporal and spatial organization of the real substrates and allowing autonomy of the new product development system.

PSM and more specifically EDOE, are promising cognitive architectures proposed as new methodologies for self-level problem solving in pharmacology.

The PSM general framework is based on four modules and their self-integrative closure.

Fig. 11.6 suggests a transition from the pharmaceutical pipelines to pipecycles.

The module K0 corresponds to substrate and resources, the module K1 to discovery step, K2 to developments and tests and the module K3 to product implementation and launching.

The first module involves resource mining. Resources are material, biological and of knowledge type.

The second module K1 is that of discovery and involves in this case drug-like molecules discovery, lead discovery and optimization. It may be a DOE.

The third module K2 is that of drug testing and development. It is a meta-design and for this reason may be denoted by 2-DOE since refers to processing DOE.

The fourth module K3 includes application and approval processes, manufacturing, marketing and monitoring of the product.

Each module may involve several sub-modules organized as epicycles.

For instance, in the module K2 there exists a natural cycle P0, P1, P2 and P3.

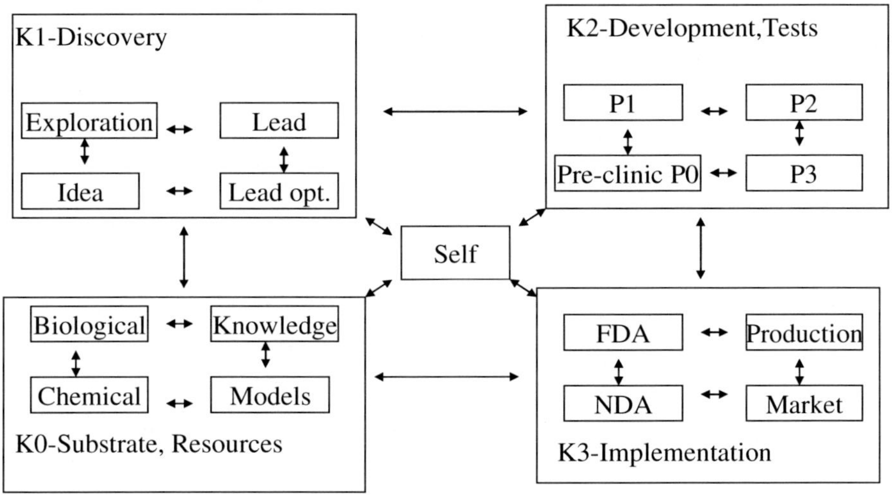

Fig. 11.6 Pharmaceutical pipecycles

For the module K3 the NDA step is followed by FDA step this by production and this by product marketing.

The transition from pipeline to pipecycles proposes a methodology that closes the loop in iterated experimentation in a high dimensional space. The cycling refers to large cycles for the whole process of four modules or just to one module or sub-module and the corresponding epicycles.

Some cycles may be fully automated if autonomous experimentation methods are used to conduct high-throughput experiments.

Modeling of matrix designs and use of informational criteria accelerate the development of new drugs.

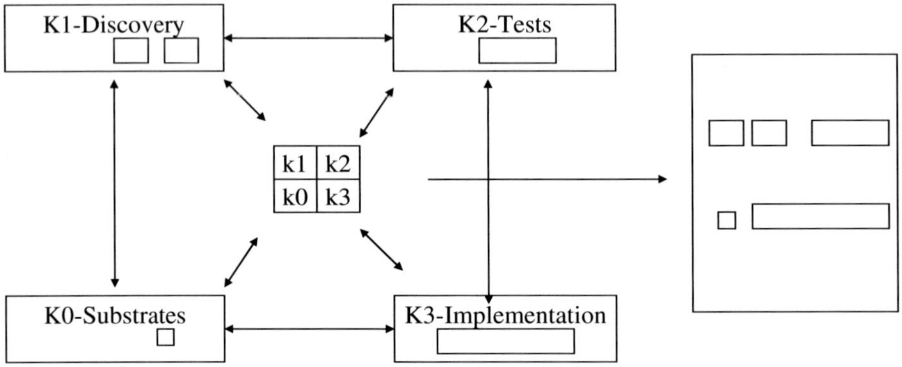

Fig. 11.7 Operad for pharmaceutical pipecycles

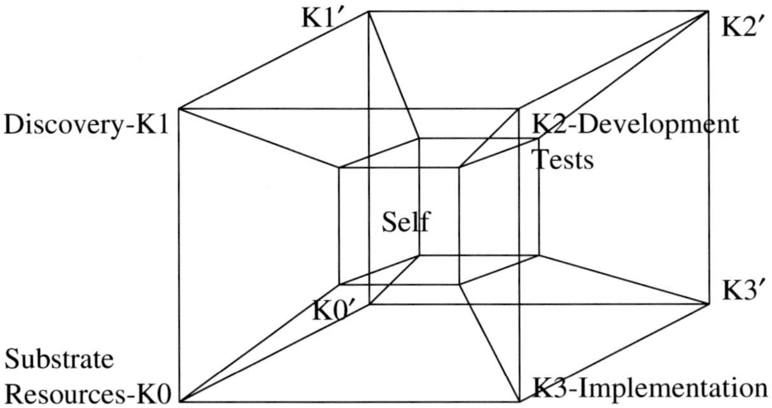

Fig. 11.8 Polytope for pharmaceutical pipecycles

Fig. 11.7 shows the operad for pharmaceutical pipecycles.

K0-Substrate, Resources, K1-Discovery, K2-Development Tests, K3-Implementation.

Fig. 11.8 illustrates the polytope of pharmaceutical pipecycles.

There exist an integration way and a differentiation way. The reconciliation of these two ways is a negotiated process since lateness in drug discovery should be avoided.

11.6 Library Design by Entropy Criteria

Applicability of PSM methodology for 2-phenylindole library design is described here.

Some 2-phenylindoles and their derivative prove to have anti-cancer activity (Basak et al. 2010).

Fig. 11.9 shows the molecular structure of the 2-phenylindole derivatives. Different radicals are denoted by R1, R2, R3 and X (see Table 1, Basak et al. 2010).

For 2-phenylindole derivatives we considered associated vectors as: y= [R1, R2, R3, X]

This means that the chosen significance order is R1>R2>R3>X.

We associate the digit "1" to R1=H, R2=H, R3= OCH_3 and X= $C(CN)_2$ and the digit "0" to the radicals that are different from these.

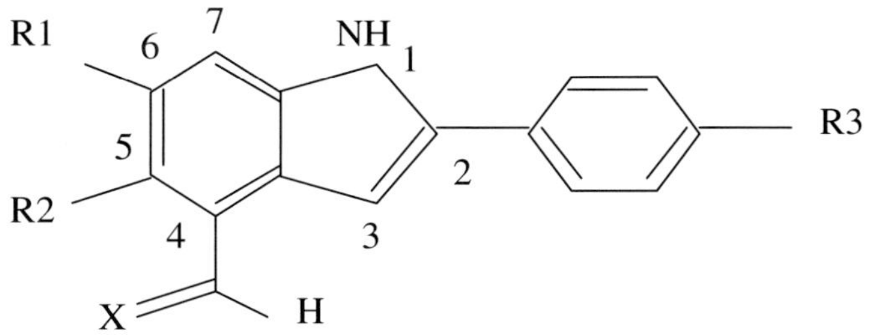

Fig. 11.9 Molecular structure of 2-phenylindole derivatives

We select the compound #2 (Table 1, Basak et al. 2010) as reference #2 = [1 1 1 1 1]

Obviously the choice of the set of characteristics and of their hierarchy should be based on drug physiological mechanism and drug efficiency tests.

Table 11.20 shows the radicals pertaining to different 2-phenylindoles derivatives.

Table 11.21 outlines the reference set for 2-phenylindole derivatives-matrix

Table 11.21 contains the same information as Table 11.20 in digitalized form.

The half maximal inhibitory concentration (IC_{50}) is a measure of the effectiveness of a compound in inhibiting biological or biochemical function. This quantitative measure indicates how much of a particular drug is needed to inhibit a given biological process by half. Here we focused on derivatives with high IC_{50}.

Table 11.20 Reference set for 2-phenylindole derivatives

No	R1	R2	R3	X	IC_{50}(nM)
#2	H	H	OCH_3	$C(CN)_2$	720
#43	H	H	H	O	420
#30	H	t-Bu	OCH_3	O	280
#19	H	n-Hexyl	CF_3	$C(CN)_2$	150

Table 11.21 Reference set for 2-phenylindole derivatives-matrix

No	R1	R2	R3	X
#2	1	1	1	1
#43	1	1	0	0
#30	1	0	1	0
#19	1	0	0	1

The compounds of the reference set have been selected to obtain a Walsh-Hadamard matrix for DOE. This offers the necessary variability for all types of drug compositions and substrate interaction.

Adding a new compound to this reference mixture, the entropy H varies.

There is only a small change of entropy, ΔH if the vector of the test compound is similar to the reference set and this supplementary compound is thought to have similar properties.

If a database 2-phenylindoles shares similar bit patterns with reference set molecules, adding a similar compound will induce a change targeting the minimum entropy production.

By contrast, inclusion of 2-phenylindoles derivatives compound having dissimilar vector leads to a higher entropy production, targeting the maximum entropy production, MEP criterion.

In this way database compounds may be screened to identify compound that causes low or high changes of the reference set informational entropy and detects other promising drugs according to the established goal.

Table 11.22 Informational entropies for 2-phenylindoles

Mixture	Reference	Reference #11	Reference #24	Reference #33	Reference #28
Matrix	[1111] [1100] [1010] [1001]	[1111] [1100] [1010] [1001] [1011]	[1111] [1100] [1010] [1001] [0110]	[1111] [1100] [1010] [1001] [0011]	[1111] [1100] [1010] [1001] [1000]
H	7.5418	11.2615	12.8343	12.8343	11.2615
ΔH	0	3.7197	5.2925	5.9525	3.7197
DD	0	0.6348	0	0	0.6348

The component #11= [1011], #24 = [0110], #33= [0011], #38= [1000] were tested since they show high (IC_{50}).

The results are shown in Table 11.22.

Table 11.22 shows the informational entropies for 2-phenylindoles.

H denotes the entropy associated to the matrix shown in first row.

ΔH denotes the difference between the entropy associated to reference and the entropy associated to reference plus one new compound.

DD denotes the distance between the reference and the matrices corresponding to reference, plus one new compound, as shown in successive columns of Table 11.22.

It appears that supplementing the reference mixture by #11, or #38, has lower effect for entropy than # 24 or #33.

The compound #11 or #38 may be preferred for a conservative new drug search based on similarity and #24 or # 33 for an innovative search based on dissimilarity.

High DD corresponds to differences allowing multiple, different classes and potential versatility of interaction. It was associated to maximum production of entropy production, MPEP criterion.

To illustrate the selection criterion at this level we take into account that the organisms varies and show biorhythms. For different regimes for organism the delivery of different 2-phenylindole mixtures may be beneficial and ensures the evolvability maximization, EM that may be evaluated by comparing DD values.

The DD criteria suggest using reference, #11 or #28, for maximum activity periods and #24 or # 33 for minimum activity periods.

EM and SEM criteria should be correlated with the methods to monitor the biomarkers of the periodic functioning of organism (Ashdown 2004, Coventry et al. 2009). Researchers have discovered that the body's immune system can destroy some cells within a window occurring every 12 to 14 days. By giving low-dose treatment at exactly the right time, they succeeded in halting the spread of advanced disease. Also they found the body has an immune cycle during which it swings "on" and "off". When the immune system turns off, it releases inhibitory cells which prevent it from fighting the disease. Treating organisms at the right time may maximize their evolvability. The timed drug delivery supposes an iterated screening of drugs and drug delivery by interaction with the organism, resources and environment.

Chronotherapy, which is an optimization of dose-time medication schedule, has been successfully applied for decades. The effects of chemotherapy exhibit circadian rhythms since the proliferation of normal cells and of damaged cells is gated by the circadian clock, damaged cells being less well synchronized. It is also known that the detoxification of cytostatic drugs depends on time of administration.

11.7 Self-Evolvable Experimentation Systems

Quantitative, predictive understanding of complex systems requires adequate information. High-throughput methods and laboratory automation technology have the potential to deliver the necessary data. To harvest this potential, experimental systems have to become evolvable and autonomous.

Self-evolvable experimentation systems are computational systems capable of autonomously investigating large experimental parameter space (Matsumaru et. al. 2004, Lovel and Zauner 2009).

Such systems should develop hypotheses, plan experiments and perform experiments in a closed-loop manner without human interaction.

Fig. 11.10 illustrates the autonomous experimentation architecture principle.

The notations are: K0-Experiment, K1-Model, K2-Prediction, and K3-Fitness.

It is a self-integrative closure technique.

The levels may be identified as follows: K0-Experiment, K1-Model, K2-Prediction, K3-Evaluation and Fitness. The model is empiric. To these the central level of self-evolution linked to the previous levels is joined. The center is considered either as the starting area or as the final area of one cycle of investigations. The swinging between the two roles should be considered too.

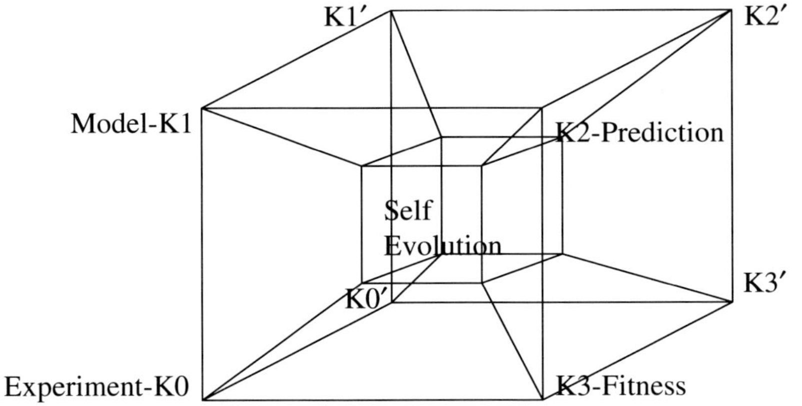

Fig. 11.10 Polytope for self-experimentation system

This suggests that after the integration or direct way we need to look at the differentiation or reverse way. Making use of the developments of the direct way will give different results for the reverse way. The swinging from direct to reverse investigation is beneficial for model evaluation and evolution.

In self-evolvable experimentation, artificial intelligence techniques are employed to carry out the entire cycle of cognition including the elaboration of hypothesis to explain observations, the design of experiments to test these hypotheses and the physical implementation of the experiments using laboratory automats to falsify hypotheses.

Investigating surprising observations, defined as those observations that disagree with a well-performing hypothesis, has been highlighted as a technique utilized by successful experimenters and has also been considered in previous computational scientific discovery techniques (Lovel et al. 2011).

A surprising observation either highlights a failure in the hypothesis or an erroneous observation. If the observation is highlighting a failure of a hypothesis, especially an otherwise well performing hypothesis with a high prior confidence, then additional experiments should be performed to further investigate the behavior where that observation was found, to allow the development of improved hypotheses. As such we consider the use of surprise to manage the direct way-reverse way trade-off, where obtaining surprising observations will lead to more direct way experiments, and unsurprising observations lead to reverse way developments.

A mathematical formulation for surprise has been considered previously in the literature is Kullback-Leibler divergence used to identify surprising improvements to the models being formed.

The informational distance DD defined by eq. 4.5 plays a similar role.

The DD may be obtained associating similarity matrices to experimental recordings and predictions (Iordache 2009).

Large value of DD states that the observation was surprising, as the overall confidence of the hypotheses has been reduced. A low value of DD states the observation was not surprising, as the overall confidence has increased. The result of DD can therefore be used to control the swinging between direct way and reverse way experiments. A large value for DD will dictate that the next experiment will be integrative, so as to allow investigation of the surprising observation. A low value of DD will lead to a reverse way experiment next, a differential one to search for new surprising features of the behavior.

Thus a hypothesis is removed from consideration if the information loss caused by the removal is small.

After DD has been calculated, the hypothesis manager will go through the process of creating new hypotheses.

This process of evaluating experiments using surprise to choose the next experiment type is automatically continued until the maximum number of experiments allowed has been performed (Lovel et al 2011).

In the coming decades a confluence of wireless networks and lab-on-chip sensor technology with application in health monitoring is expected. In such lab-on-chip network each sensor node is endowed with a limited supply of chemicals. The network will collectively or via the self-evolution level decide how the drug resources will be spent.

Environmental monitoring and improving new drugs and new material discoveries may be performed by similar autonomous experimentation architectures.

References

Ashdown, M.L.: Timed ablation of regulatory CD4$^+$ T cells can prevent murine AIDS progression. J. Immunol. 172, 4917–4925 (2004)

Bailey, R.A.: Efficient semi-Latin squares. Statistica Sinica 2, 413–437 (1992)

Basak, C.S., Zhu, Q., Mills, D.: Prediction of anticancer activiyty of 2-phenylindoles: Comparative molecular field analysis versus ridge regression using mathemathical descriptors. Acta Chim. Slov. 57, 541–550 (2010)

Cioppa, T.M., Lucas, T.W.: Efficient Nearly Orthogonal and Space-filling Latin Hypercubes. Technometrics 49(1), 45–55 (2007)

Coventry, B.J., Ashdown, M.L., Quinn, M.A., Markovic, S.N., Yatomi-Clarke, S.L., Robinson, A.P.: CRP identifies homeostatic immune oscillations in cancer patients: a potential treatment targeting tool? Journal of Translational Medicine 7(102), 1–8 (2009)

Hedayat, A.S., Sloane, N.J.A., Stufken, J.: Orthogonal Arrays. Theory and Applications. Springer, New York (1999)

Iordache, O.: Evolvable Designs of Experiments Applications for Circuits. J. Wiley VCH, Weinheim (2009)

Iordache, O.: Polystochastic Models for Complexity. Springer, Heidelberg (2010)

Kolda, T.G., Bader, B.W.: Tensor decompositions and applications. SIAM Review 51(3), 455–500 (2009)

Lohr, S.L.: Hasse Diagram in Statistical Consulting and Teaching. The American Statistician 39(4), 376–381 (1995)

Lovell, C., Jones, G., Gunn, S., Zauner, K.-P.: Autonomous Experimentation: Active Learning for Enzyme Response Characterisation. In: JMLR: Workshop and Conference Proceedings, vol. 16, pp. 141–155 (2011)

Lovel, C.J., Zauner, K.-P.: Towards algorithms for autonomous experimentation. In: Naef, F., Galliot, B., Nehaniv, C. (eds.) Proceedings of the 8th International Conference on Information Processing in Cells and Tissues (IPCAT 2009), pp. 150–152 (2009)

Matsumaru, N., Centler, F., Zauner, K.-P., Dittrich, P.: Self-Adaptive Scouting—Autonomous Experimentation for Systems Biology. In: Raidl, G.R., Cagnoni, S., Branke, J., Corne, D.W., Drechsler, R., Jin, Y., Johnson, C.G., Machado, P., Marchiori, E., Rothlauf, F., Smith, G.D., Squillero, G. (eds.) EvoWorkshops 2004. LNCS (LNAI), vol. 3005, pp. 52–62. Springer, Heidelberg (2004)

Nguyen, N.-K.: A new class of orthogonal Latin hypercubes (2008)

Taguchi, G.: Introduction to Quality Engineering: Design Quality into Products and Processes Tokyo. Asian Productivity Organization (1986)

Taguchi, G.: Systems of Experimental Design, vol. 1 & 2. Unipub/Kraus International, White Plains (1987)

Wood, J., Rose, V.S.: Method of designing chemical substances. PCT WO 99/26901 (1999)

Woodcock, J., Woosley, R.: The FDA critical path initiative and its influence on new drug development. Annu. Rev. Med. 59, 1–12 (2008)

Ye, K.Q.: Orthogonal Latin hypercubes and their application in computer experiments. J. Amer. Statist. Assoc. 93, 1430–1439 (1998)

Chapter 12
Perspectives

Abstract. Selfdisciplinarity is presented as a necessary step in problem solving for evergrowing complexity systems.

Answering to the demand for systems able to combine technologies, sciences, and engineering into condensed expressions, the polytope project is proposed. This project starts from a general architecture shared by the operational structure of self-evolvable devices, the functional organization of organisms as informational and cognitive systems, and the scientific and engineering methods.

Conceptual, selfware, hardware, fabrication and applications perspectives of this project are sketched.

12.1 Selfdisciplinarity

The domain of evergrowing complexity concerns the problems that can be seen in the nature, industry and society and are considered as very hard or intractable.

These include problems like traffic control, diseases as pandemic influenza, genetic drugs design, cognitive architectures, control and manufacturing systems, environment data and experiment organization, market evolution and so on. We tend to throw up our hands at these problems, thinking that individually, we cannot make a difference, or that the problems are just too complicated.

What these problems all have in common, actually, is that they exhibit a hierarchy of emergent patterns caused by the local and global interactions of a large number of individual agents. We lack the scientific tools to think consistently about such problems (Conklin 2006).

It has been argued in many ways that the problem solving for high complexity domain is an activity which cannot succeed on the basis of one point of view, or the knowledge of one discipline, but that it needs cooperation of a number of disciplines to develop valid knowledge.

Confronted with an explosion of new disciplinary knowledge, it is difficult for any specialist to understand more than a fraction of his specialized domain. The management of the cooperation of different disciplines for complex problem solving is a concern. Consequently, it is necessary to find ways to radically simplify and unify knowledge about complexity.

Piaget and Garcia methodology starts from the hypothesis that there exists a parallelism between the particular problem solving and the historical development of the involved sciences (Piaget and Garcia 1989). The short history of an individual problem solving, that is the problem ontogeny, is considered as parallel to the evolutionary long history of a lineage that is, the problem phylogeny. The isomorphism between psychogenesis and the historical development in sciences is explained by the general equilibration based on assimilation accommodation mechanism and instantiated as the so-called intra-inter-trans process.

The intra-inter-trans process is the functional mechanism that proceeds from simple object analysis, the so-called *intra* step, to the analysis of relations between objects via transformations, that is the *inter* step, and to the building of cognitive structures, that is the *trans* step.

This general mechanism is relevant to both particular problem solving and to scientific activity itself. Piaget considered that the general intellectual development involves the same sequence of steps. In particular, he reconstructs development from sensory-motor stage, to pre-operational thinking corresponding to the *intra* stage, via concrete-operational thinking corresponding to the *inter* stage, toward formal-operational thinking, that corresponds to the *trans* stage. In a larger Piagetian view, the claim is that this kind of stage can be traced in different domains and at all levels of development.

The intradisciplinarity step is linked to single disciplinarity or to multidisciplinarity realm. It represents the first step of the problem solving.

Disciplinary research is not able to fully cope with complex problems because these problems do not fit into the conventional system of scientific disciplines. Complex problems referring to energy, food and drugs, health, ecology, security and financial problems cannot be solved by disciplinary approaches. A scientific understanding of complex problems is mandatory but the increasing specialization and fragmentation of scientific disciplines prevents disciplinary research from working.

Multidisciplinarity makes use of different disciplines and suppose that studying complex problem is not just in one discipline only, but in several, at the same time. Any issue in question will be enriched by incorporating the perspectives of several disciplines.

Multidisciplinary approach brings a benefit to the disciplinary study, but this benefit is still in the restricted service of the source disciplines. The multidisciplinary approach runs over disciplinary boundaries while its goal remains limited to the frameworks of disciplinary research.

The next step to be considered in problem solving methodology is that of interdisciplinarity. This involves cooperating disciplines and has a different goal than multidisciplinarity. It concerns the transfer of methods from one discipline to another. Like multidisciplinarity, the interdisciplinarity spreads out the disciplines.

The next step in complex problem solving is the transdisciplinarity. The definition of problems to solve is, for this step, relatively independent of disciplinary perspectives. Transdisciplinarity concerns that which is at once between the disciplines, across the different disciplines and beyond disciplines (Nicolescu 2002, 2006).

Growing complexity problems do not belong to only one of the three main types or disciplinarity sketched above but contain elements of each type.

Fig. 12.1 illustrates the problem-solving polytope.

The environment contains the real data and conventional methods.

Initially these parts are separated but start to form well-defined disciplines in the first stage, K1, that is, the *intra* stage. They may be coupled in the second stage, K2, that is, the *inter* stage to form interacting disciplines. The third stage, K3, the *trans* stage corresponds to the coupling of two or more sciences in wide-ranging frameworks, avoiding disciplinary isolation and going beyond disciplines.

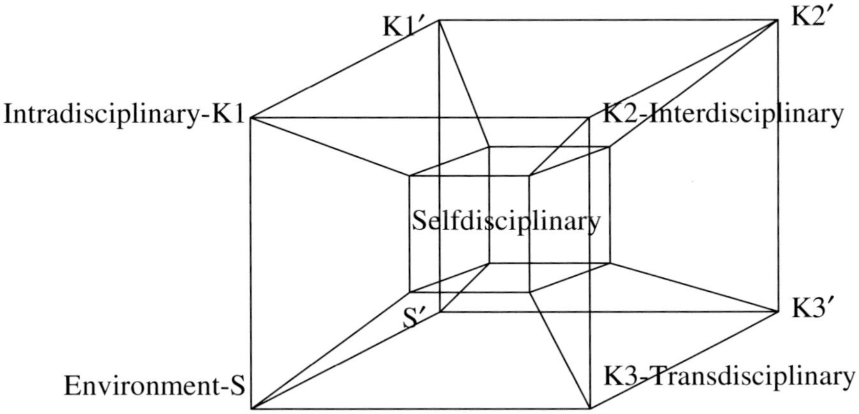

Fig. 12.1 Polytope for selfdisciplinarity

The fourth stage, shown in Fig. 12.1, may represent an integrative or the Self viewpoint. After a complete cycle intra-inter-transdisciplinarity, the Self viewpoint is open toward a new disciplinary approach and a new cycle. This fourth stage completes and recombines the knowledge cycle and the problem solving. It corresponds to the post-formal or creative stages in development and supposes the ability to formulate post-disciplinary notions as for instance new axioms and new goals.

Selfdisciplinarity joints recent trends advocating the convergence of several disciplines as, nanoscience, biotechnology, information technology and cognitive science known as the NBIC concept (Bainbridge and Roco 2006). Convergence is a new paradigm that can yield critical advances in a broad array of sectors, from health care to energy, food, and climate (Sharp et al. 2011, Sharp and Langer 2011).

The overarching request correlates selfdisciplinarity to the concept of metadisciplinarity (Scott and Shurville 2005, von Stillfried 2007).

By metadisciplinarity we mean a discipline about disciplines. It comments on the forms and procedures that constitute particular disciplines. A component of metadisciplinarity is that it brings to completion the transdisciplinary endeavor of uniting all disciplinary perspectives but also uniting the disciplinary with interdisciplinary and transdisciplinary approach.

Metadisciplinarity actually not only points to the place from where everything started but at the same time sets the stage for a whole new level of differentiation and integration, by opening and closing the circular pattern into a kind of spiral movement.

A particular view of the selfdisciplinarity polytope is the Piaget's cycle of sciences that includes S-Sciences of matter (physics, chemistry), K1-Biosciences (biology, anatomy), K2-Cognitive sciences (psychology, sociology), K3-Mathematics and Logics. This cycle was described by Piaget (Piaget 1967).

Fig. 12.2 shows the polytope of scientific disciplines.

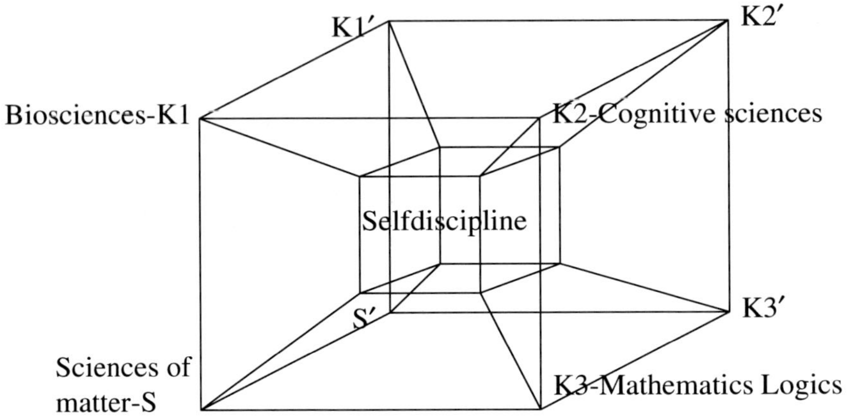

Fig. 12.2 Polytope for scientific disciplines

Selfdisciplinarity refers to a new dimension but how the integrative or selfdisciplinary viewpoint turns back into a new disciplinary life is an open problem. The selfdisciplinary process leading to the formulation of a new understanding and possible new discipline is as important as the resulting understanding itself. A suggestion is that self-evolvable problem solving may restart and follow the same steps on a higher methodological plane that is at a higher dimension in modeling. This means that the architecture of the Self inner cube parallels that of the external cube in figures as Fig. 12.2.

Following categorification way, a decategorification way should be considered too.

Fig. 12.1 and Fig. 12.2 show that after the integration, or direct epistemology way S→K1→K2→K3 we need to look at the differentiation, or reverse epistemology way K3′→K2′→K1′→S′.

This kind of reverse epistemology was studied by Bailly for the cycle of disciplines (Bailly 2010). It was observed that making use of the developments of the direct way will offer in a kind of symmetry-breaking result. On account of this, the swinging from direct to reverse epistemology will be beneficial since creative and new information supposes coexistence of integration and differentiation.

For the polytopes of sciences, it should be observed that any new step of the cycle embeds elements of the previous ones. The higher order should be inclusive and self-aware on previous levels. After the integrative closure, the material embodiment of logics, mathematics and computing capacity will allow operating the material realm at multiple levels simultaneously. This may support the emergence of another type of sciences of matter of biosciences, of cognitive sciences and so on. Consequently a spiral of sciences instead of cycle of sciences and associated systems may be taken into account as a more appropriate image of knowledge development (Iordache 2009, 2010).

This spiral image has been discussed by several authors (von Stielfield 2007, Bailly et al. 2010). Through the spiral shape of time circles can be fully interpreted. The spiral image suggests that history of knowledge is never repeated. It is just similar, as identical events happen, but always under different circumstances.

Finally let us observe that selfdisciplinarity refer to research and problem solving that combines disciplines that are already known as related, as for instance design and engineering. This links the selfdisciplinarity to already existing polytechnic disciplinarity. Without doubt, the connection between theory and practice that is between K3 and S levels is mandatory for engineers. Selfdisciplinarity is project based and it demonstrates an ability to pound together ideas, disciplinary problem and to create new ways of working, new practices, unexpected processes and engineering projects.

12.2 The Glass Bead Game

Complexity is the research field emerging around the conviction that some problems of organization in domains as material science, molecular biochemistry, neuroscience, computer science, telecommunications, manufacturing and economy can be challenged scientifically in a unified way, by means of which progress in understanding aspects of organization in either field can be fruitful to the others. By integrating disparate fields, we may link very different disciplines that can learn and benefit from one another.

The process of finding unifying principles either at the microscopic or macroscopic levels of complex systems is hindered both by the divisions between specialized disciplines and by the problems of technical language where different concepts share overloaded names while similar concepts may have different names (Buchli and Santini 2005).

Despite substantial knowledge about complex systems, the application of this knowledge to the engineering domain remains difficult. Efforts to manage complexity are scattered over many scientific and engineering disciplines.

Attempts to establish complexity engineering as a discipline are hindered by misunderstandings over basic terms such as emergence and causation. It is improbable that the consensus making will be successful while more disagreements complicate the use of common terms (Haken 1999). Although terminology standardization is a necessary feature of communication, it can also pose a barrier impeding the technological progress.

Standard should be at the same time flexible and rigorous.

As the amount of knowledge keeps growing exponentially and the subject areas we deal with are getting exceedingly complicated, more concentrated, if possible minimal ways of conveying knowledge should be developed and implemented.

Herman Hesse's novel, *The Glass Bead Game*, suggests some symbolic ways to confront high complexity in the 21st century (Hesse 1969).

Hesse envisages the glass bead game, GBG, as a system that has been able to combine technologies, sciences, philosophy and mathematics into one condensed expression, a new language with a new grammar.

Cast in a future period, one in which intelligent activity reached its broad expression the GBG provided a way for researchers and engineers from various disciplines to synthesize their thinking into new planes of knowledge (DeLisi 1999).

The GBG, imagined by Hesse, resembled an abacus, or in modern words a computer or a self-reconfigurable automaton, with several dozen wires strung vertically, horizontally or along the edges. Upon these wires were hung beads of various colors, sizes and shapes. Moving the beads into new configurations symbolically represented the development of new themes. Over time, the GBG was adopted by all major disciplines with the beads corresponding to the artifacts, symbols, formulae and notations of that respective discipline.

As the GBG developed over many years, it became increasingly desirable to develop the common language and grammar across several fields in order to make their similarities and differences clearer and to build an institution around the development, preservation and communication of this new language and paradigm of global culture.

The GBG would transcend different disciplines and allow researchers from these disciplines to interact, and hence, to learn from one another.

The same capability to build upon each other's ideas is described in the current studies devoted to higher complexity. Such studies describe the transdisciplinary and selfdisciplinary work of researchers in the fields of technology, biology, economics, information sciences and physics, and describe how new insights, for example in market study, emerge from thinking in the field of molecular biology. This may be an example of GBG in action.

An ultimate illustration of the search for GBG is the mathematical categorification.

By categorification one can understand, very generally, presenting a notion in a categorical setting, which usually involves generalizing the notion and making advanced distinctions.

In the context of mathematics, the beads of GBG corresponded to mathematical formulae and theorems, which were combined with the mathematical notations of other players, to form new insights. The same mathematical structure has many different empirical realizations since a mathematical domain deals with more than one empirical context. This relationship between mathematics and the external world suggests a similar relation between category theory and mathematics. All the mathematical fields can be organized according to their structure by specific categories, and such specific categories can be organized using the notion of general category as provided in category theory. Category theory is a general formalism, but there is a certain special way that mathematicians, physicists and engineers use categories which turns out to have close analog in different domains as topology, logic, computation, and so on (Baez and Stay 2008).

Mathematical categorification is the process of finding category-theoretic analogs of set-theoretic concepts by replacing elements with objects, sets with categories and so on. The term categorification refers also to the process in which ordinary categories are replaced by the n-categories. In higher dimensional category theory researchers encounter a ladder which they are irresistibly drawn to ascend, step by step, from 0-categories to 1-categories, to 2-categories and so on (Baez and Dolan 1998, Corfield 2005).

This ladder proves to be a polytope, since the ascending portions are tied to descending ones as in the coupled categorification and decategorification processes.

12.3 Polytope Project

12.3.1 Concepts and IT

The polytope project for a biologically inspired multi-purpose architecture, useful for artifacts building, information representation, designs, operations and calculus, is presented here.

The project assigns the polytopic character in the way we are looking for necessary messages into essential objects that can be seen from many different perspectives.

Reflecting different aspects, physical, technological, scientific and socio-economical, the resulting architectures will be also interesting in themselves as geometrical objects like n-cubes, lattices and polytopes.

The issues raised by this project concern the foundational machine structure, the hardware and software, the scientific and engineering methods.

The project is based on findings from material science and electronics, biology, psychology and informatics and it is expected to provide a general framework for subsequent quantitative and theoretical research in these domains.

Projects having in part similar objectives pertain to the field of high dimensional automata, OLAP project (Berson and Smith 1997), cgmCUBE project (Dehne et al. 2006), CUBIST project (Dau 2011), programmable matter and self-reconfiguration of modular robots project (Goldstein et al. 2005, Gilpin and Rus 2010) and so on.

Similar objectives can be detected for biologically inspired computing initiatives such as natural computing (de Castro 2006), autonomic computing (Kephart and Chess 2003) and organic computing (Würtz 2008).

The polytope project encompasses conceptual and IT, architectural and application aspects.

We start by discussing conceptual and IT aspects.

Constantly growing amounts of data complicated and rapidly changing interactions, and an emerging trend of incorporating unstructured data into analytics, are bringing new challenges to conventional IT and computing devices.

Current solutions involve IT users dealing with increasingly complex systems analyses.

But conventional system programming paradigms, investigation methods and management tools are not designed for handling the scale, the growing complexity, or the dynamism and heterogeneity of emerging network and systems.

Biosystems have developed strategies to cope with dynamic, complex, highly uncertain constraints. For this reason modern research area of IT tried to apply biosystems concepts to solve its unsolved problems related to high complexity.

A significant objective is to dispose and manipulate information in a condensed and significant form.

Looking to biosystems for inspiration we will discuss two already related aspects, the categorification and the semantic capabilities (Cockett 2006).

IT solutions have neglected the categorical aspects of data and models, and this can be the source for uncontrolled and unsafe behavior. It is the case of some high dimensional automata (Fajstrup et al. 2006). Several critical problems for automata safe behavior have been discussed by Bringsjord in relation to categorification (Bringsjord et al. 2010).

It was observed that automata need logical system that includes not only deontic operators, but also epistemic operators for beliefs and knows and a full calculus for time, change, goals, and plans.

Moreover automata need to solve software verification problems and need to take account of the fact that reasoning ranges over many different kinds of logical systems, and involves integrative meta-reasoning of the systems.

Ethical reasoning, like reasoning in the formal sciences, finally sends to the Piaget's post formal stages and to the problem of conscious machines (Haikonen 2007).

The proposed solution for the control of automata in high complexity environments should be based on categorification.

Categorification process allows significant data gathering.

Categorification consists in regulating the behavior of automata with specific codes rendered in computational logic, so that all actions they perform are provably permissible relative to these codes. One promising approach to elaborate this formally is the n-category theory, where categories are logical systems.

It is expected that human-like cognition, whether or not it is directed by specified categorical codes, exploits coordinated functors over many logical systems encoded as categories. These systems range from the propositional calculus, through description logics, to first-order logic, to temporal, epistemic, deontological, and so on.

Cognitive systems operate in ways that range across a large number of logical systems. So, the polytope project needs to develop a formal theory, and a corresponding set of processes that captures the meta-coordination of several logical systems. This relates the project to the domain of linear logic and polycategories (Cockett 2006).

Categorized technologies, focusing on the meaning of data, should be able of dealing with both unstructured and structured data. Having the meaning of data and a categorical reasoning mechanism in place, a user can be better guided during an analysis.

The challenge is to develop IT methods, including bridges between real systems, and category concepts like the categorical imperative, codes and so on.

Conventional IT solutions neglected also the semantics or in other words, the meaning of data, which can limit the completeness of analysis and make it difficult. For example to remove redundant data coming from different sources, we need meaning to confront redundancy. A piece of information can be semantically selected and explained or a new relevant fact can bring to the user's attention.

Semantic analysis will improve classical methods in IT, such as data reduction and duplicate detection.

In particular, it is expected that semantic techniques as the evolvable DOE, EDOE and the lattices as studied in Formal Concept Analysis, FCA, will be key elements of new IT systems.

EDOE represents a modern way to replace pre-programmed and fixed problem-solving methods by flexible and self-evolvable ones. EDOE allows directing, focusing and rationalizing the data acquisition and interpretation (Iordache 2009).

Concept lattices have been studied as categories. Moreover, results in lattice theory may be a source of inspiration for category theory too.

It should be observed that semantic technologies as EDOE or FCA have traditionally operated on small data sets if compared to classical IT developments.

The polytope project should develop methodologies and a platform that combines essential features of categorized, semantic technologies and IT.

The critical problems for the polytope project consist in identifying the dual ways in the polytopic frame, the Self exemplars and the synchronization rhythms of the dual ways to be considered for specific problems and systems.

Dual pairs are those things, events and processes that are mutually related and inextricably connected. Such dualities are dynamic and relational. Both aspects of a dual pair are required for an exhaustive account of phenomena (Engstrom and Kelso 2008).

The inspiration for dualities comes from the study of complementarities in physics and of duality in mathematics. The inspiration comes also from cognitive systems that are working by such dualities.

This refers to the biology of the human brain, namely, the dual nature of the hemispheric specializations to the dual nature of brain processes and explains how is the brain functionally organized to achieve self-adaptive behavior in a changing world.

A promising choice for the dual ways in complex problem solving may be the pair design and lattice. We may consider EDOE and FCA as example of pair.

Following EDOE step, the FCA step should be considered and so on.

EDOE implementation is followed by data acquisition and representation as FCA.

This may modify the structure of DOE giving rise to a new FCA and so on.

This swinging between designs EDOE for data acquisition and data representation FCA, empowers both these coexisting methods and allows data understanding.

Swinging between FCA and EDOE can be used to guide a user in discovering new facts, which are not explicitly modeled by the initial data storing schemas.

The project involves self-evolvability capability for both EDOE and FCA and for the whole dual system. This concerns the Self capability.

To identify Self exemplars and to understand how the Self drives the dual ways to confront complexity are critical problems.

For the Self-understanding and building we need to look for inspiration to ribosomes, neocortex role, neuronal global workspace, post-formal cognitive stages, to core arrays in HCA, and to antipodes in Hopf algebras.

The Self should be able to mediate and to correlate the dual ways.

It is the right rhythm and interaction of both ways that counts for self-evolvability.

Inspiration for rhythms comes from synergetics in physics, meta-stability in neuroscience or biorhythms and chronotherapies in biology and pharmacology.

The problem is that one needs to identify beforehand the rhythms whose utilization may be beneficial or detrimental for the particular system.

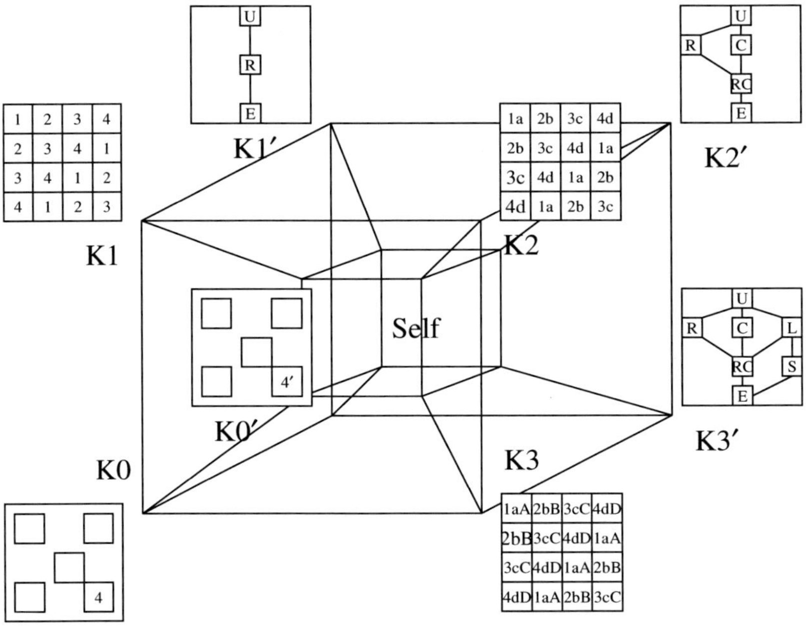

Fig. 12.3 Duality EDOE and Hasse Diagrams

Fig 12.3 shows an illustrative example of duality between EDOE frame and the lattice associated to the statistical analysis of the data analyzed by that design.

The DOE is based on semi-Latin squares (Bailey 1992).

It is illustrated on the front face of the outer cube in Fig. 12.3.

The module K0 contains unstructured items, data and information.

The module K1 is a DOE organizing the objects 1, 2, 3 and 4 as a Latin square.

The module K2 adds attributes a, b, c and d to the objects.

The module K3 continues to associates the conditions A, B, C and D.

Instead of these designs based on Latin squares we can consider simpler designs containing only "0" and "1". It is the case of Walsh-Hadamard designs.

It is known that every locally finite poset has a naturally associated Hasse diagram.

The Hasse diagram associated to the semi-Latin design is shown on the back face of the outer cube in Fig. 12.3.

Let Ω be the set of n^2k points which are divided into n rows and n columns in a way that the intersection of each row with each column contains k points.

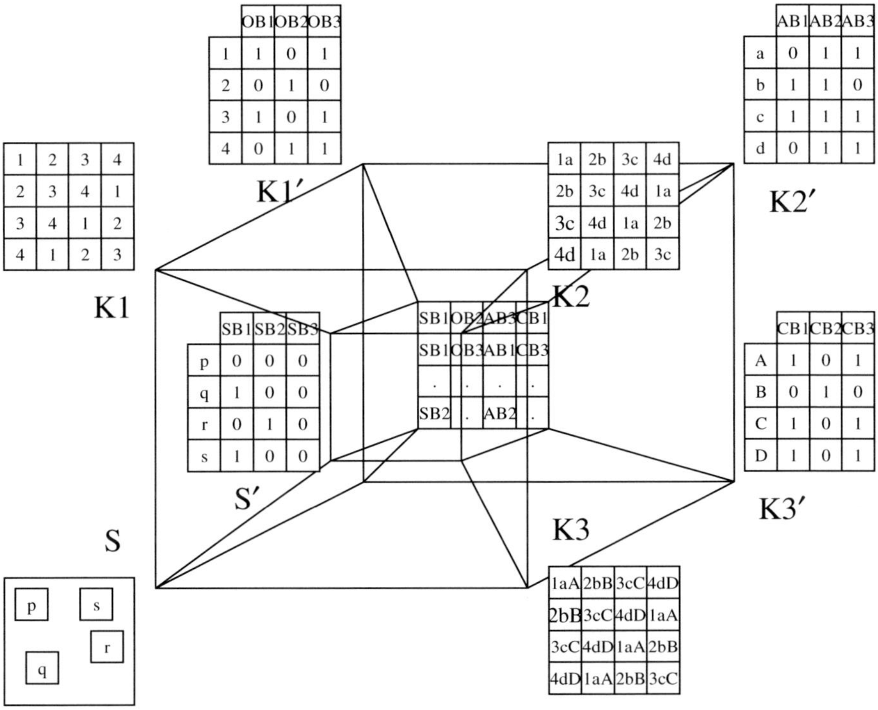

Fig. 12.4 Duality EDOE and HCA

R, C, S denote the partition of Ω into rows, columns and symbols. RC=R $\vee$ C and L= (R $\wedge$ C) $\vee$ S. Here $\vee$ denotes the supremum and $\wedge$ denotes the infimum. E denotes the partition of Ω in n^2k singletons and U denotes the trivial partition of Ω containing a single class. U is called universal factor while E is called the equality factor (Bailey 1992).

Level after level the design inflates adding new letter in DOE matrices.

The levels are taken into account in the associated Hasse diagram.

These refer to rows R for K1′, to rows R, columns C and their interaction RC for K2′. The symbols S are added for K3′. It is the natural construction in triadic FCA, objects, attributes and symbols as conditions.

A statistical analysis method as ANOVA shows if new factors or interactions should be taken into account (Lohr 1995). An example shown in Fig. 12.3 is the object 4′ in K0′.

This is a modified object 4 from K0.

Fig. 12.4 shows an example of duality between an evolvable DOE and HCA frame.

DOE modules are represented on the front face of the outer cube from Fig. 12.4.

The module S refers to substances as s, p, q, and r, to unstructured objects, data and so on. The module K1 is a 1-DOE organizing the objects 1, 2, 3 and 4.

The module K2 associates attributes a, b, c and d to the objects.

The module K3 continues to associate the conditions A, B, C and D.

HCA modules are represented on the back face of the external cube shown in Fig. 12.4.

This refers to bundles as substance bundles, SB for S′, objects bundles OB for K1′, attribute bundles AB for K2′ and condition bundles CB for K3′.

The Self involves the elements of the core array developed in HCA methods. Core array indicates the linking structure among the hierarchies.

Notice that instead of the designs based on Latin squares for EDOE we can consider simpler designs containing only “0” and “1” as shown by HCA method.

On the front face of the outer cube of the polytope, we have actively imposed matrices of design, while on the back face we have passively recorded matrices of data.

12.3.2 Architecture

Examples of basic polytopic architectures are shown in Fig.12.5 and Fig. 12.6

Fig. 12.5 is based on the 4-cube. Cubelets are present in all corners of the inner and outer cube but they may fill the vertices and the inner spaces too.

The cubelets are supposed to receive information, analogical or digital and transfer this.

Moving the cubelets into new configurations symbolically represented the solutions or development of new problems. Swinging between different faces of the outer and inner cubes allows gaining information from direct way and reverse way in investigation.

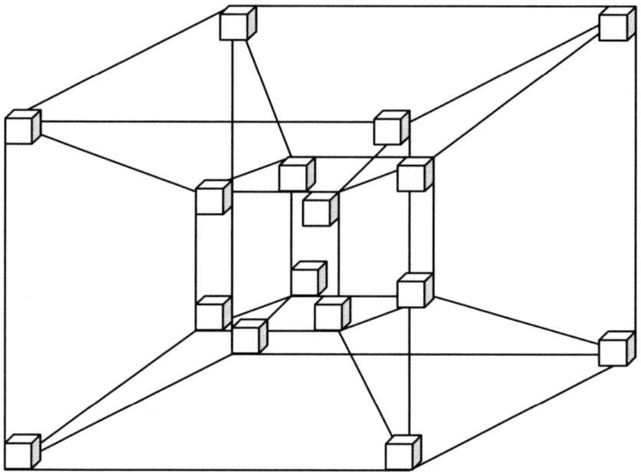

Fig. 12.5 Polytope based on 4-cube

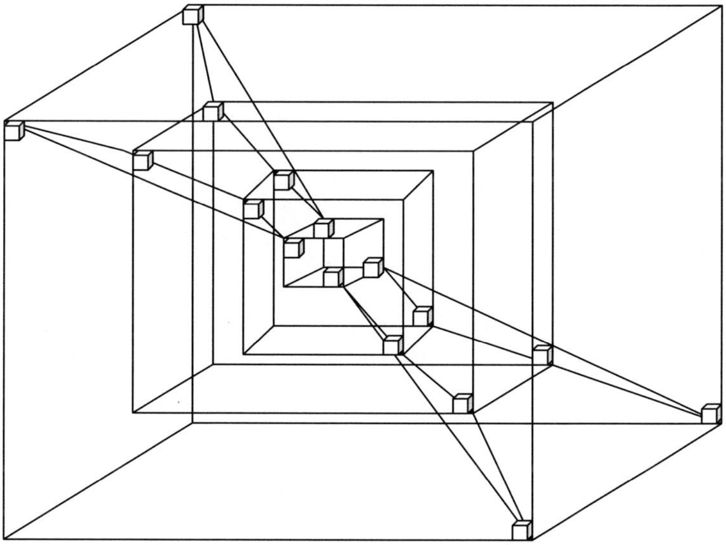

Fig. 12.6 Polytope based on 5-cube

The polytope shown in Fig. 12.6 is based on a 5-cube (Joswig and Ziegler 2000).

For comprehensibility reasons only a part of the 5-cube is represented and decorated with cubelets. The 5-cube potentialities for investigation are dramatically increased if compared to 4-cube architectures. A hierarchy of meta-stability domains, Self modules and rhythms should be considered.

A challenge when building with discrete modules as cubelets, pebbles or beads is that the designers must simultaneously reconcile the shape and the behavior of the architecture. Fig. 12.7 shows a hierarchical organization (Fig. 12.7a) and a modular organization (Fig. 12.7b).

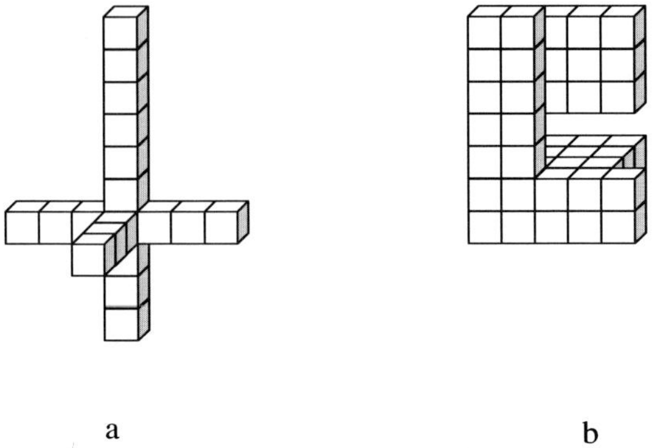

Fig. 12.7 Hierarchy and modularity

To address the concerns related to design of architectures, it is necessary to develop algorithms that can control the shape without detailed extensive planning or communication. We need to allow basic planning and significant communications.

A collection of pebbles or cubelets as shown in Fig. 12.7 can be viewed as a kind of programmable matter (Goldstein et al. 2005, Gilpin and Rus 2010, Schweikhard and Gross 2011).

Architectures based on fine-grained modular automata represent a platform for self-evolvable systems.

This addresses the design, fabrication, motion planning, and control of autonomous kinematical machines with variable morphology. Beyond conventional actuation, sensing, and control typically found in fixed-morphology robots, self-reconfigurable robots are also able to deliberately change their own shape by rearranging the connectivity of their parts in order to adapt to new circumstances, perform new tasks, or recover from damage.

One can imagine large numbers of tiny cubic robotic modules, working together to create larger polytopic tools, devices, automata and so on.

In contrast to large, expensive and complex automata, self-evolvable automata systems show polytopic architectures of identical modules which can be programmed to assemble themselves in multiple configurations for multiple tasks.

Rather than deploy a family of fragile, custom-made architectures and automata, a polytope of modules, pebbles, or cubelets, could be delivered, configuring themselves as necessary, self-organizing, planning and communicating, self-repairing and so on.

Among the benefits of modular self-evolvable polytopes we may consider versatility, reliability, resilience and cost.

While specific large automata created for a specific task are often suited only to that task, polytopic reconfigurable automata should be able to adapt to different tasks in different environments. Large automata may be expensive, and often unreliable, while small modules organized in polytopic frames can be mass-produced for vast cost savings.

Most of the existing designs are based on homogeneous modules that is, cubelets of identical components which connect with each other to form the polytopic assembly.

In real field conditions, heterogeneous systems will dominate. This follows from the fact that useful automata need many specialized parts, including specific sensors, actuators and effectors tools corresponding to cubelets of different types. Including every part and function in every tiny module is expensive, so modules of various types will be included in a self-evolvable system. In addition, when self-reconfigurable robots are further miniaturized, fewer components can be included in each module, so the resulting heterogeneity must increase.

To develop the mathematical models used for engineering design of the polytope projects is a challenging task.

Over the past century the most fundamental tools for engineers have been differentiation, integration and differential models. These models allow the detailed design of artifacts.

For polytope project we need new type of models that will allow now to design the shapes, the architectures without details, schemas, experimental designs and so on.

It was observed that the new types of models are formally similar to the classical ones.

They capture the intuitions from the ordinary calculus since we have calculus rules of differentiation and integration expressed algebraically, formally similar to the classical ones.

The EDOE are based on models as wave equation, WE, model.

Schemas and circuits may be based on differential categories.

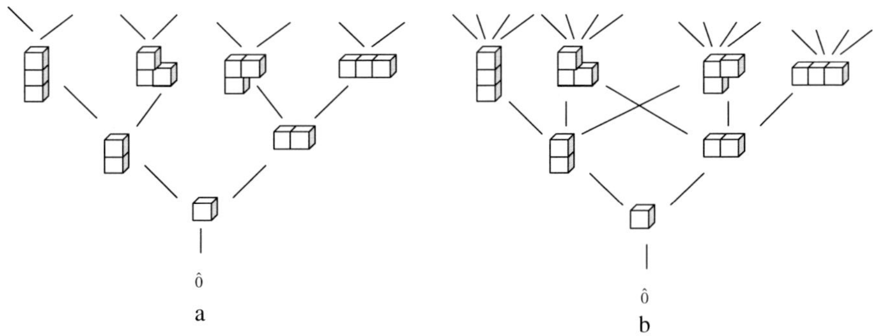

Fig. 12.8 Dual constructions

Lattices represent an important part of the polytope project. Lattice theory refers to posets and Hasse diagrams, to developments as FCA and HCA. Differential posets are the models generating lattices.

A large variety of polytopes may be generated by the differential models as wave equation WE, differential posets or differential categories.

Young lattices are among of the most studied differential posets.

Fig. 12.8 shows an example of dual constructions, the dual graded graphs for the Hecke algebra (Bergeron et al. 2011).

A self-evolvable construction should be able to swing between them.

Reconfigurable tableau or lattices may be built by cube-style modular robots (Aloupis et al. 2009, Gilpin and Rus 2010).

To correlate the micro-automata reconfigurability domain with the differential posets formalism is an exciting task. We refer to lattices and dual graded graphs as resulting by Robinson–Schensted–Knuth, RSK-algorithms.

The differential model expresses the rules to build the polytope.

Fig. 12.9 illustrates the process of forming shapes through assembly and disassembly.

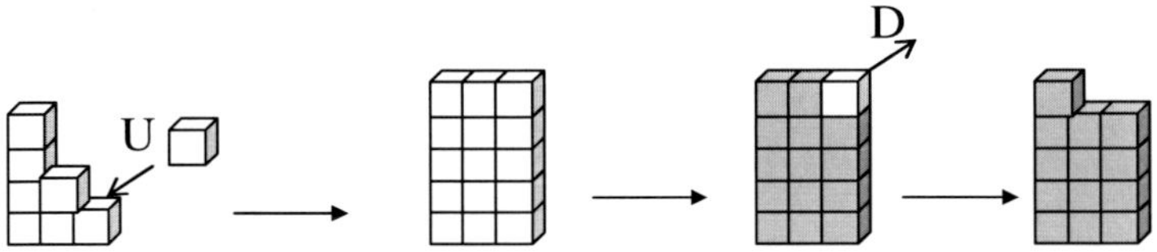

Fig. 12.9 Assembly and disassembly

Initially a regular block of material results by modules assembly characterized by the operator U. Once this material structure is completed, the modules not needed in the final structure detach from the neighbors. The process is described by the operator D.

Once these extra modules are removed, we are left with the final shape.

The process is that governed by the operators U and D for differential posets.

Assembly and disassembly are dual concepts, also in a categorical sense and need dual algebras for modeling.

The magnification and shape duplication corresponds to doubling and contracting operations as described for lattices. This modeling tool was applied in robotics for modular shape magnification (An and Rus 2010).

The polytopes should be able to perform operations as: addition of new elements to have a word or string, modification in interior of a chain by small cycle performing, rotation and change line in column.

They may contain Latin squares and cubes, semi-Latins and Walsh-Hadamard functions.

All these prove to be solutions of the differential models.

Polytopes may have a fractal structure and will contain filled and void areas.

It is a need for new ways of applying visualization tools in which meaningful diagrammatic polytope representations will be used for data depicting, for navigating through the data and for visually querying the data.

FCA and EDOE may benefit from existing IT functionalities such as OLAP.

OLAP synthesis can combine the methods developed in self-reconfiguring robots (An and Rus 2010) with that developed in the study of relational OLAP (Chen et al 2004).

In this case the modules to be assembled are the processors.

An illustration of the potentialities is offered by the process of browsing the data cube (Han et al 2011).

Fig. 12.10 illustrates the browsing process.

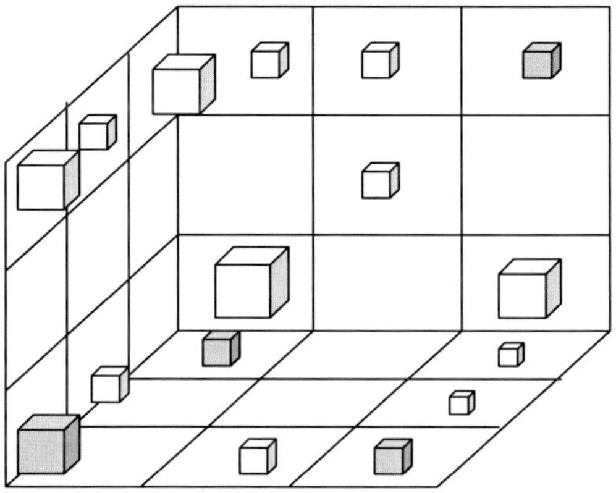

Fig. 12.10 Browsing data cube

This process allows visualization, focusing and interactive manipulation at both hardware and selfware levels.

The process is similar to magnification or duplication process in self-configurating automata.

In the same time OLAP operations as drill-up and down, slice and dice, rotate and drill across or drill through, may be introduced in the micro-automata program.

12.3.3 Applications

Practical implementations of the polytope project are self-evolvable separation schemas.

This refers to dual separation schemas, duality in cyclic operations, reconfigurable separation schemas.

As the self-evolvable circuits we refer to: polytopes as antennas, solar cells, batteries, patches and so on.

Antennas for instance may contain polytopes, cubes, and cubelets and be able to detect non-standard signals. To fabricate such devices we need to use existing printed circuits, patches fabrication methods and 3-D technologies as molding.

Manufacturing may implies new methodologies as self-evolvable manufacturing system represents an implementation of self-evolvable FCA and EDOE concepts in manufacture organization.

The main objective of the polytope project for biological systems is to understand and to make use of similar architectures as suggestion for artificial systems.

To re-apply this understanding to find new explanations of biological relevance for real biosystems may be considered as a long-term objective only.

This concerns bio-inspired computers, cognitive and control architectures.

The existing self-configurating automata are based on macroscopic elements, in the best cases millimetric ranges. For lower range we need to consider devices based on biological materials as substrata. We refer to bacterio-rhodopsine layers cubes or polytopes and *Physarum*-based polytopes.

The process may be continued at molecular level too (Nagpal 2002, Whiteside, Grzybowski, 2002, de Castro 2006).

The project will be a support for coagulation of data from a variety of unstructured and structured real sources. It would enable a user to perform IT operations over semantic and categorized data. It will help to develop autonomous semantic and categorized automata in hospitals, personalized drug design, drug delivery and health care.

The project should demonstrate the resulting technology progress in the fields of scientific data acquisition analysis, computational biology, market intelligence and the field of control center operations.

Other areas for future research are, traffic control, visualization, meteorology, environment, ecology, energy management, cars and homes personalized architecture, market and so on.

References

Aloupis, G., Collette, S., Damian, M., Demaine, E.D., Flatland, R., Langerman, S., O'Rourke, J., Ramaswami, S., Sacristan, V., Wuhrer, S.: Linear reconfiguration of cube-style modular robots. Computational Geometry - Theory and Applications 42, 652–663 (2009)

An, B., Rus, D.: Making Shapes from Modules by Magnification. In: IEEE/RSJ International Conference on Intelligent Robots and Systems (2010)

Baez, J.C., Dolan, J.: Categorification. In: Getzler, E., Kapranov, M. (eds.) Higher Category Theory, Contemp. Math., vol. 230, pp. 1–36. American Mathematical Society (1998)

Baez, J., Stay, M.: Physics, Topology, Logic and Computation: A Rosetta Stone. In: Coecke, B. (ed.) New Structure for Physics. Lecture Notes in Physics. Springer (2008)

Bailey, R.A.: Efficient semi-Latin squares. Statistica Sinica 2, 413–437 (1992)

Bailly, F.: L'anneau des disciplines. Enquête sur quelques concepts théoriques et gnoséologigues. AFSCET, Paris (2010)

Bailly, F., Longo, G., Montévil, M.: A 2-dimensional geometry for biological time. In: Biologiee Selezioni Naturali Conference, Florence, December 4-8, 2009 (2010)

Bainbridge, W.S., Roco, M.C. (eds.): Managing Nano-Bio-Info-Cogno Innovations: Converging Technologies in Society. Springer Science and Business Media, Berlin (2006)

Bergeron, N., Lam, T., Li, H.: Combinatorial Hopf algebras and Towers of Algebras – Dimension. Quantization and Functorality, arXiv:0710.3744v1 (2011)

Berson, A., Smith, S.J.: Data Warehousing, Data Mining, and OLAP. McGraw-Hill (1997)

Bringsjord, S., Taylor, J., Wojtowicz, R., Arkoudas, K., van Heuvlen, B.: Piagetian Roboethics via Category Theory: Moving Beyond Mere Formal Operations to Engineer Robots Whose Decisions are Guaranteed to be Ethically Correct. In: Anderson, M., Anderson, S. (eds.) Machine Ethics. Cambridge University Press, Cambridge (2010)

Buchli, J., Santini, C.: Complexity engineering, harnessing emergent phenomena as opportunities for engineering. Tech. Rep. Santa Fe Institute Complex Systems Summer School, NM, USA (2005)

Chen, Y., Dehne, F., Eavis, T., Rau-Chaplin, A.: Parallel ROLAP data cube construction on shared nothing multiprocessors. Distributed and Parallel Databases 15, 219–236 (2004)

Cockett, J.R.B.: What is a good process semantics? (2006), `http://pages.cpsc.ucalgary.ca/~robin/talks/estonia.pdf`

Conklin, J.: Dialogue mapping: Building shared understanding of wicked problems. John Wiley & Sons, Chichester (2006)

Corfield, D.: Categorification as a Heuristic Device. In: Gillies, D., Cellucci, C. (eds.) Mathematical Reasoning and Heuristics. King's College Publications (2005)

Dau, F. (ed.): Proceedings of the 1st CUBIST Workshop 2011. CEUR-WS, vol. 753 (2011)

de Castro, L.N.: Fundamentals of Natural Computing: Basic Concepts, Algorithms, and Applications. CRC Press (2006)

Dehne, F., Eavis, T., Rau-Chaplin, A.: The cgmCUBE project : Optimizing parallel data cube generation for ROLAP. Distrib. Parralel Databases 19, 29–62 (2006)

DeLisi, P.S.: The Glass bead Game Linking Interdependence and organizational learning (1999), `http://www.org-synergies.com/docs/Glass-Bead-Game.pdf`

Engstrom, D., Kelso, J.: Coordination dynamics of the complementary nature. Gestalt Theory 30(2), 121–134 (2008)

Fajstrup, L., Goubault, E., Raussen, M.: Algebraic topology and concurrency. Theoret. Comput. Sci. 357(1-3), 241–278 (2006)

Gilpin, K., Rus, D.: Modular Robot Systems: From Self-Assembly to Self-Disassembly. IEEE Robotics and Automation Magazine 17(3), 38–53 (2010)

Goldstein, S.C., Campbell, J.D., Mowry, T.C.: Programmable matter. IEEE Comput. 38(6), 99–101 (2005)

Haikonen, P.O.: Robot Brains: Circuits and Systems for Conscious Machines. Wiley & Sons, Chichester (2007)

Haken, H.: Information and Self-Organization A Macroscopic Approach to Complex Systems. Springer, Berlin (1999)

Han, J., Kamber, M., Pei, J.: Data Mining: Concepts and Techniques, 3rd edn. Morgan Kaufmann (2011)

Hesse, H.: The Glass Bead Game. Holt, Rinehart and Winston, Inc., New York (1969)

Iordache, O.: Evolvable Designs of Experiments Applications for Circuits. J. Wiley VCH, Weinheim (2009)

Iordache, O.: Polystochastic Models for Complexity. Springer, Heidelberg (2010)

Joswig, M., Ziegler, G.M.: A neighborly cubical 4-polytope. Electronic Geometry Models, No. 2000.05.003, C45 Master.poly. (2000)

Kephart, J.O., Chess, D.M.: The vision of autonomic computing. IEEE Computer 36(1), 41–50 (2003)

Nagpal, R.: Programmable self-assembly using biologically-inspired multiagent control. In: Proceedings of the 1st International Joint Conference on Autonomous Agents and Multi-Agent Systems (AAMAS), pp. 418–425. ACM Press, New York (2002)

Nicolescu, B.: Manifesto of Transdisciplinarity. SUNY Press, New York (2002)

Nicolescu, B.: Transdisciplinarity-Past, Present and Future. In: Haverkort, B., Reijntjes, C. (eds.) Moving Worldviews - Reshaping Sciences, Policies and Practices for Endogenous Sustainable Development, pp. 142–166. COMPAS Editions, Holland (2006)

Piaget, J.: Classification des sciences et principaux courants épistémologiques contemporains. In: Piaget, J. (ed.) Logique et Connaissance Scientifique, Gallimard, Paris, pp. 1151–1224 (1967)

Piaget, J., Garcia, R.: Psychogenesis and the History of Science. Columbia University Press, New York (1989)

Schweikardt, E., Gross, M.D.: Experiments in design synthesis when behaviour is determined by shape. Pers Ubiquit. Comput. 13, 123–132 (2011)

Scott, B., Shurville, S.: Epistemological Unification of the Disciplines: The Contributions of Socio-Cybernetics. In: Proceedings of the Sixth European Congress on Systems Science, Paris, France (2005)

Sharp, P.A., Cooney, C.L., Kastner, M.A., Lees, J., Sasisekharan, R., Yaffee, M.A., Bahatia, S.N., Jacks, T.E., Lauffenburger, D.A., Langer, R., Hammond, P.T., Sur, M.: The Third Revolution: The Convergence of the Life Sciences, Physical Sciences, and Engineering. MIT White Paper (2011)

Sharp, P.A., Langer, R.: Promoting Convergence in Biomedical Science. Science 333, 527 (2011)

von Stillfried, N.: What about Transdisciplinarity? Its Past, its Present, its Potential... and a Proposal. In: Transdisciplinarity and the Unity of Knowledge: Beyond the Science and Religion Dialog, Philadelphia, Pennsylvania (2007)

Whitesides, G.M., Grzybowski, B.: Self-assembly at all scales. Science 295, 2418–2421 (2002)

Würtz, R.P. (ed.): Organic Computing: Series: Understanding Complex Systems. Springer (2008)

Appendices

Abstract. Modeling at several abstraction levels, as required by self-evolvable systems, is based on higher dimensional category theory.

Category theory concepts as operads, rewriting, and polycategories are presented as abstract counterparts of operation schemas, calculus and polystochastic models.

Appendix 1 n-Categories

A category is specified by a class of objects and a class of arrows called also morphisms.

In category theory, the objects or identity arrows are elements within category, whereas the category compares objects, the functors compares categories and the natural transformation compares functors.

The n-categories are high-order generalizations of the notion of category (Leinster 2004).

An n-category is the algebraic structure consisting of a collection of objects, a collection of morphisms between objects, a collection of 2-morphisms between morphisms and so on up to n, with various rational ways of composing theses j-morphisms, $j<n$.

The 0-category is a set, while 1-category is a category. An n-category consists of 0-cells (objects, types), 1-cells (morphisms), 2-cells (morphisms between morphisms) and so on, all the way up to n-cells together with composition operations.

As n increases, the construction of n-categories step by step may be difficult to conceive and need analysis on how the n-categories are effectively working.

Consider for example the case of 2-categories of which the category of categories denoted by Cat, is the standard example (MacLane 1971). In Cat, the 0-cells are categories, the 1-cells are functors, and the 2-cells should be natural transformations.

Any 2-category C makes use of three items C_0 , C_1 , and C_2. Elements of C_i are called i-cells i =0, 1 or 2. The 2-category is the three categories structure that consists of the base category having C_0 as objects and C_1 as arrows, the horizontal category having C_0 as objects and C_2 as arrows, and the vertical category having C_1 as objects and C_2 as arrows.

The 2-cells are arrows in both the horizontal and the vertical category, thus they composes with two different composition operators, horizontal or vertical.

Cat is a strict 2-category, that is, all laws hold exactly, not just up to isomorphism.

Vertical composition corresponds to a sequential operation, while horizontal composition corresponds to a parallel operation. The 2-category is a category with morphisms between morphisms, that is, 2-morphisms.

There are also many weak categories. For example a bicategory is a notion used to extend the notion of 2-category to handle the cases where the composition of morphisms is not strictly associative, but only associative up to an isomorphism.

Bicategories may be considered as result of the weakening of the definition of 2-categories. A similar process for 3-categories leads to tricategories, and more generally to weak n-categories for n-categories.

A tricategory C is defined informally by:

- A class C_0 of objects
- For any pair A, B $\in C_0$ a bicategory C (A, B)
- For any triplet A, B, D$\in C_0$ a bifunctor of composition c_{ABD}: C(A, B) x C(B, D)$\rightarrow$ C(A, D)
- For any object a bifunctor u_A: 1 $\rightarrow$ C (A, A)

These elements verify several axioms (Gordon et al. 1995).

By categorification one can understand, very generally, presenting a notion in a categorical setting, which usually involves generalizing the notion and making more distinctions. More specifically, categorification is the process of finding category-theoretic analogs of set-theoretic concepts by replacing elements with objects, sets with categories, functions with functors and equations between functions by natural isomorphisms between functors, which in turn should satisfy certain equations of their own, called coherence laws (MacLane 1971, Baez and Dolan 1998).

The term coherence covers in category theory what from a logical point of view would be called problems of completeness, axiomatizability and decidability. An influential notion of coherence is due to MacLane (MacLane 1963). At the same time when coherence started being investigated in category theory, the connection between category theory and logic was established.

One may categorify an algebraic structure by taking its objects to be arrows of a category. The notion of category is a categorification in this sense of the notion of monoid, the monoids being categories with a single object. The motivation for categorification may be internal to category theory, but it may come from other areas of mathematics, like algebraic topology and mathematical physics lin particular, quantum field theory (Baez and Dolan 1998).

The correspondence between set theory and category theory is presented in Table A1.1

Decategorification is the reverse process of categorification. Decategorification is a systematic process by which isomorphic objects in a category are identified as equal._Categorification is more complicated than decategorification, and requires insight into individual situations.

Table A1.1 Correspondence between sets and categories

Set theory	Category theory
Set elements	Objects
Sets	Categories
Functions	Functors
Equalities between morphisms	Natural isomorphisms of functors

The term vertical categorification refers roughly to a process in which ordinary categories are replaced by n-categories. Categorification implies moving from left to right in the periodic table of categories while decategorification implies moving in the reverse direction (Baez and Dolan 1998, Leinster 2004).

Appendix 2 Operads

Originating from work in algebraic topology by Boardman and Vogt and May operads have recently found many applications (Boardman and Vogt 1973, May 1972).

An operad is an abstraction of a family of composable functions of n-variable for various n, useful for accounting and applications of such families.

A nonsymmetrical topological operad is a sequence of topological spaces A_0, A_1, A_2, together with unit $e \in A_1$ and a multiplication map:

μ: $A_k \times A_{i1} \times A_{i2} \times \ldots \times A_{ik} \rightarrow A_{i1+i2+..+ik}$

This satisfies associativity and unitarity condition.

Here A_k indicates the way to put together, while A_{i1}, …, A_{ik} indicates things to put together.

The nonsymmetrical operads form a category. A significant example is represented by the Stasheff associahedra, K_n (Stasheff 1963a, b)

The associahedra K_4 is shown in Fig.A.1. If we fix a specific associating homotopy and consider the five ways of parenthesizing the product of four loops, it results a pentagon whose edges corresponds to a path of loops (Fig.A.1).

The associahedron K_n can be described as a convex polytope with one vertex for each way of associating n ordered variables, that is, ways of inserting parentheses in a meaningful way in a word of n letters. The edges correspond to a single application of an associating homotopy.

The sequence of associahedra $\{K_n\}$ forms an operad, the A_∞ -operad (Stasheff 1963a, b).

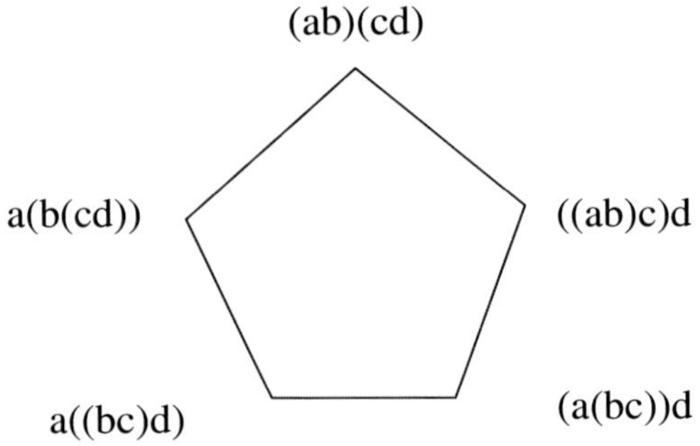

Fig. A.1 Associahedra K_4

The little cubes operads is another significant example of operads (Boardman and Vogt 1973).

Fig.A.2 illustrates this operad.

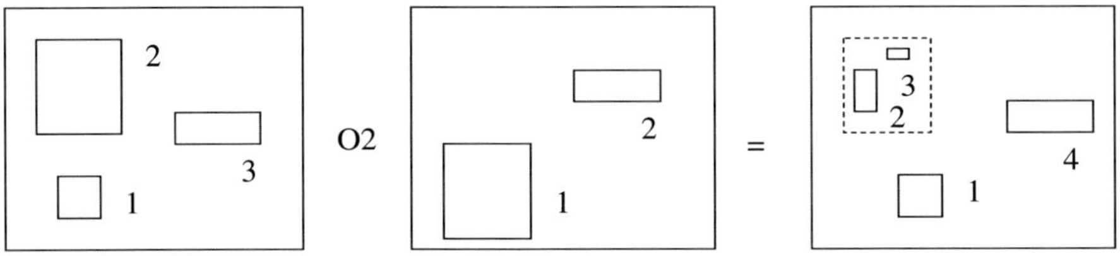

Fig. A.2 Little 2-cubes operad

The algebra over an operad A is a topological space X together with a map of operads:

A→ End (X)

Operads in a category of topological spaces represents the synthesis of several approaches to the recognition problem for iterated loop spaces. Beginning with Stasheff associahedra and Boardman and Vogt little n-cubes and continuing with Batanin's approach to n-categories through higher operads that problem has extensively been studied (Batanin 1998, 2003).

An advance is based on the connection between iterated monoidal categories and iterated loop spaces (Balteanu et al. 2003, Forcey et al. 2007).

Appendix 3 Rewriting

In computer science, mathematics and logic, rewriting covers a wide range of methods of replacing sub-terms of a formula with other terms. Rewriting systems are also known as reduction systems. In their most basic form, they consist of a set of objects, plus relations on how to transform those objects.

One rule to rewrite a term could be applied in many different ways to that term, or more than one rule could be applicable. Rewriting systems then do not provide an algorithm for changing one term to another, but a set of possible rule applications. When combined with an appropriate algorithm, however, rewrite systems can be viewed as computer programs, and several declarative programming languages are based on term rewriting.

Rewriting systems have broad applications in general model of computation, Petri nets, for automated theorem proving and so on.

Rewriting rules specify the repeated replacement of sub-terms of a given formula with equivalent terms. A rewriting rule is a basic derivation that allows passing from one term in an appropriate language to another and the study of a rewriting system is the study of the compositions of such basic derivations. Higher dimensional categories naturally appear in the study of various rewriting systems.

The string rewriting systems and the categorical notation for such systems are introduced following Johnson (Johnson 1991). If S is a set that is an alphabet, let us denote S* the free monoid on S. This means that the elements of S* are words in the alphabet S.

The string rewriting system consists of an alphabet S and a set R$\subset$ S* x S*. We write an element (s, s′) $\in$ R as s→s′ and call it a rewrite of s to s′. An element w$\in$ S* may be rewritten by finding a sub word of w, witch match the left hand side of a rewrite and replacing it with the corresponding right hand side to obtain some w′. We could write w→w′ and say that w and w′ are → related. A sequence of such rewrites w→w′→…→w″ is called a derivation.

Observe that the random systems with complete connections as defined by Iosifescu and Theodorescu could be considered as a string writing system a kind of reverse process for rewriting (Iosifescu and Theodorescu 1990).

Consider for example S= {a,b,c,x,y,z} and R={xyz→a, ab→x}. An example of derivation beginning with w=xyzbyabc is: xyzbyzabc→abyzabc→xyzabc→ xyzxc→axc.

The application of the rewrite ab→x in the context abc→xc could have been carried out at any position in the sequence of rewrites. Thus a distinct but essentially equivalent derivation would be: xyzbyzabc→abyzabc→abyzxc→xyzxc→axc

The rewrite taking abc→xc may be thought of as occurring in parallel with the other rewrites.

This can be made explicit in a 2-categorical perspective. The 2-categories provide a simple framework for string rewriting systems. A 2-category consists of objects indicated as points, arrows called 1-cells which go between objects and which compose in the usual way, and arrows called 2-cells which go between 1-cells and which compose horizontally or vertically. The 1-cells describe relations while the 2-cells describe relations between relations. Horizontal composition corresponds to parallel operations while vertical composition corresponds to sequential operations. In a 2-category all compositions are associative, there are identity 1-cells and 2-cells and the horizontal and vertical compositions interact according to the interchange law which say that composing first horizontally and then vertically gives the same result as composing first vertically and then horizontally. Observe that a 2-category C is a Cat-enriched category.

The linear rewriting schemas may be studied as 3-categories.

The 2-categories have objects, morphisms, 2-morphisms and some axioms while the 3-categories have objects, morphisms and 3-morphisms together with some axioms.

If the structures of interest are 2-categories the computations in these structures are 3-categories. A 3-category is a 2-Cat enriched category. In this case there are arrows called 3-cells which go between 2-cells. The 3-cells are generated by the rewrite schema.

The 3-categorical formulation of term rewriting systems allows an axiomatization of rewriting and critical pair completion in terms of patterns, (certain specified 2-cells), multipliers, (2-categorical composition) and replacements, (certain specified 3-cells) (Buchberger 1987, Johnson 1991).

Appendix 4 Polycategories

Detailed formal presentations of polycategories can be found in literature (Pastro 2004, Koslowski 2005).

Informally a planar polycategory P consists of the following data:

- A class P_0 of objects of P,
- For each m; n ∈ N and $X_1,\ldots,X_m$, $Y_1,\ldots,Y_n \in P_0$, a set $P(X_1,\ldots,X_m, Y_1,\ldots,Y_n)$ whose elements are called polymorphisms. Using Γ and Δ to represent strings of elements of P_0, the polymorphisms in P (Γ; Δ) may be denoted

  $f: \Gamma \rightarrow \Delta$ or $\Gamma \vdash_f \Delta$ where dom (f) = Γ and cod (f) = Δ
- For each $X \in P_0$, an identity morphism $1_X \in P(X; X)$
- An operation

$P(\Gamma; \Delta_1, X, \Delta_2) \times P(\Gamma_1, X, \Gamma_2; \Delta) \rightarrow P(\Gamma_1, \Gamma, \Gamma_2 ; \Delta_1, \Delta, \Delta_2)$ called cut, restricted to the cases where either Γ_1 or Δ_1 is empty and either Γ_2 or Δ_2 is empty. This restriction is called the crossing or planarity condition.

Planarity condition on composition leads to four possible cases where composition may occur. Composition of processes is modeled by the cut rules (Cockett 2006).

Fig.A.3 is an illustration of the cut rule.

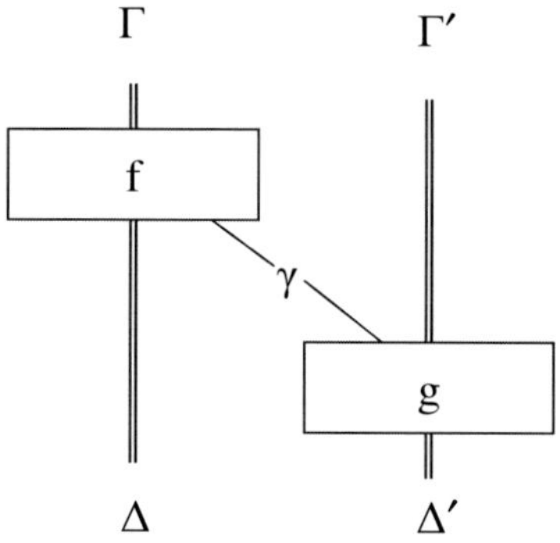

Fig. A.3 Cut rule

In this case the processes f and g are composed via the protocol rule denoted by γ.

Composition must satisfy the interchange and associative laws.

Interchange laws are concerned with diagram as shown in Fig.A.4, and associativity laws with diagrams as shown in Fig.A.5.

In these cases, f, g and h represent processes, and α, β and γ represents protocols.

A polycategory is symmetric in case $P(\sigma\Gamma; \tau\Delta) = P(\Gamma; \Delta)$ for permutations σ and τ and coherence conditions hold. This is stating that the order of the objects coming into a process and coming out of a process is not important. The symmetry map is composable and must commute with cut. This means that when polycategories are symmetric, crossing pipes or wires are allowed.

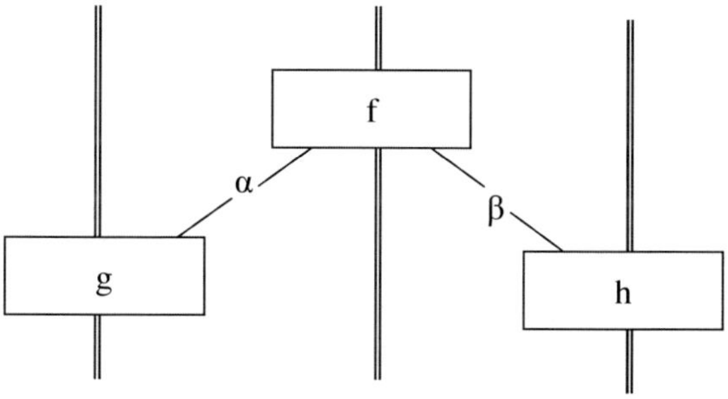

Fig. A.4 Interchange

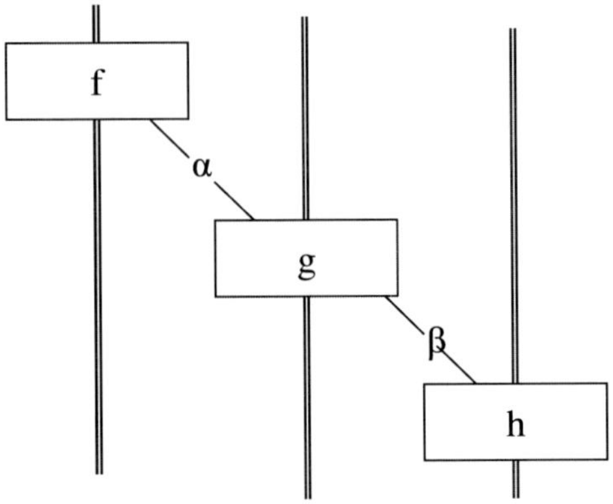

Fig. A.5 Associativity

Schemas and circuit diagrams represent accepted ways of visualizing polycategories.

Notice that this is the opposite from usual drawing explaining categories were the maps are drawn as arrows or connectors and the objects as boxes.

An important characteristic of polycategories is representability (Cockett 2006).

A polycategory is representable in case there are polynatural equivalences:

$r_{\otimes}$: P $(\Gamma_1, X, Y, \Gamma_2 ; \Delta) \rightarrow$ P $(\Gamma_1, X \otimes Y, \Gamma_2 ; \Delta)$
$r_{\top}$: P $(\Gamma_1, \Gamma_2 ; \Delta) \rightarrow$ P $(\Gamma_1, \top, \Gamma_2 ; \Delta)$
$r_{\oplus}$: P $(\Gamma; \Delta_1, X, Y, \Delta_2) \rightarrow$ P $(\Gamma; \Delta_1, X \oplus Y, \Delta_2)$
$r\bot$: P $(\Gamma_1, \Gamma_2 ; \Delta) \rightarrow$ P $(\Gamma_1, \bot, \Gamma_2 ; \Delta)$

These results by replacing commas with composite types: $\otimes$, $\top$, $\oplus$, $\bot$

Polynatural means that the transformation is invariant under cutting into the non-active position.

It was established that representable categories corresponds to linearly distributed categories (Blute et al. 1996, Cockett and Seely 1997).

This means that in a representable polycategory:

- The composite types: $\otimes$, $\top$ and $\oplus$, $\bot$ become monoidal structure
- There are coherent natural transformations
  δ_L: $A \otimes (B \oplus C)\ ) \rightarrow (A \otimes B) \oplus C$
  δ_R: $(B \oplus C) \oplus A \rightarrow B \oplus (C \otimes A$

These are called the linear distributions.

Linearly distributive categories are the appropriate framework for considering a specific logical system known as linear logic, introduced by Girard (Girard 1987).

A distributive lattice with $\wedge = \otimes$ and $\vee = \oplus$ is an example of linear distributive category.

A distributive category is a linearly distributive category with respect to the product and coproduct if and only if it is a poset.

References

Baez, J.C., Dolan, J.: Categorification. In: Getzler, E., Kapranov, M. (eds.) Higher Category Theory, Contemp. Math. 230, pp. 1–36. American Mathematical Society (1998)

Balteanu, C., Fiedorowicz, Z., Schwänzl, R., Vogt, R.: Iterated Monoidal Categories, Adv. Adv. Math. 176, 277–349 (2003)

Batanin, M.: Monoidal globular categories as a natural environment for the theory of weak n-categories, Advances in Mathematics. Advances in Mathematics 136, 39–103 (1998)

Batanin, M.: The combinatorics of iterated loop spaces, arXiv: math.CT/0301221 (2003)

Blute, R., Cockett, J.R.B., Seely, R.A.G., Trimble, T.: Natural deduction and coherence for weakly distributive categories. Journal of Pure and Applied Algebra 113, 229–296 (1996)

Boardman, J.M., Vogt, R.M.: Homotopy Invariant Algebraic Structures on Topological Spaces. Lecture Notes in Mathematics, vol. 347. Springer, Berlin (1973)

Buchberger, B.: History and basic features for the critical pair/completion procedure. J. Symbolic Computation 3, 33–38 (1987)

Cockett, J.R.B.: What is a good process semantics? (2006), `http://pages.cpsc.ucalgary.ca/~robin/talks/estonia.pdf`

Cockett, J.R.B., Seely, R.A.G.: Weakly distributive categories. J. Pure Appl. Algebra 114, 133–173 (1997)

Girard, J.-Y.: Linear logic. Theoretical Computer Science 50, 1–102 (1987)

Forcey, S., Siehler, J., Seth Sowers, E.: Operads in iterated monoidal categories. Journal of Homotopy and Related Structures 2, 1–43 (2007)

Gordon, R., Power, A.J., Street, R.: Coherence for tricategories. Memoirs Amer. Math. Soc. 117(558) (1995)

Iosifescu, M., Grigorescu, S.: Dependence with complete connections and applications. Cambridge Univ. Press, Cambridge (1990)

Johnson, M.: Linear term rewriting systems are higher dimensional string rewriting systems. In: Rattray, C.M.I., Clark, R.G. (eds.) The Unified Computation Laboratory, pp. 101–110. Oxford University Press (1991)

Koslowski, J.: A monadic approach to polycategories. Theory and Applications of Categories 14, 125–156 (2005)

Leinster, T.: Higher Operads, Higher Categories. Cambridge University Press (2004)

Mac Lane, S.: Natural associativity and commutativity. Rice University Studies. Papers in Mathematics 49, 28–46 (1963)

MacLane, S.: Categories for the Working Mathematician. Springer, New York (1971)

May, P.: The Geometry of Iterated Loop Spaces. Lecture Notes in Math., vol. 271. Springer, Berlin (1972)

Pastro, C.A.: ΣΠ-polycategories, additive linear logic, and process semantics. Master's thesis, University of Calgary (2004)

Stasheff, J.D.: Homotopy associativity of H-spaces, I. Trans. Amer. Math. Soc. 108, 275–292 (1963a)

Stasheff, J.D.: Homotopy associativity of H-spaces, II. Trans. Amer. Math. Soc. 108, 293–312 (1963b)

Self-Evolvable Systems

Octavian Iordache

Self-evolvability solutions to ever growing complexity should be envisaged giving that conventional engineering and scientific methodologies and technologies based on learning and adaptability come to their limits and new ways are nowadays required.

Self-evolvability paradigm refers to self-organizing, self-managing, self-repairing and other capabilities of the systems able to act on their own behalf in a challenging environment.

Closure and categories, polystochastic models, differential models, and informational entropy criteria are introduced in the chapters 2, 3 and 4 dedicated to methods. Chapter 5 outlines self-evolvability for physical and chemical systems as separation schemas, growing dendrites and circuits. Chapter 6 analyzes major biochemical models and bio-inspired computing initiatives. Chapter 7 outlines self-evolvability features of different cognitive and agent based architectures. Chapter 8 focuses on self-evolvable control capability and architectures. Chapter 9 is dedicated to self-evolvable manufacturing systems. Concept lattices, multivariate analysis and applications for biochemical systems are presented in Chapter 10. Chapter 11 evaluates self-evolvability potential for designs of experiments, with applications to pharmaceutical systems and quality evaluations. Chapter 12 evaluates the perspectives and proposes the polytope project. Introducing a new field of major practical and theoretical interest the book will be a valuable reference for engineers, scientists, entrepreneurs and students managing the necessary transition from learning and adaptability to self-evolvability for technologies, devices, products, problem-solving methods and organizations.

Keywords: bio-inspired computing, categories, circuits, cognitive architecture, differential models, dendrites, design of experiment, duality, entropy, hypercube, hypercycle, lattice, manufacturing, operad, polytope, polystochastic, schemas, self-evolvable, self-properties, wave equation

Index

H

I

K

L

M

N

O

MIX
Papier aus verantwortungsvollen Quellen
Paper from responsible sources
FSC® C105338

Printed by Books on Demand, Germany